CONRAD, THE LATER MORALIST

by

John E. Saveson

RODOPI N.V.
AMSTERDAM 1974

Library of Congress Catalog Card Number: 74–82789
ISBN: 90–6203–248–6

ACKNOWLEDGMENTS

The following study was made possible by summer fellowships in 1970 and 1973 awarded by the Newberry Library (Chicago) and by a sabbatical leave in 1973-74 granted by Mansfield State College (Pennsylvania). I am indebted to Dr. Lawrence Towner, Director, and Dr. James Wells, Associate Librarian, at the Newberry; and to Dr. Lawrence Park, President of the Mansfield State College.

Some parts of this study have appeared in *Conradiana,* 6 (1974), 57-62; *Costerus Essays in English and American Language and Literature,* 8 (1973), 177-92; *Criticism,* 14 (1972), 32-48; *Modern Language Studies,* 2 (1972), 59-62; *Review of English Studies,* 24 (1973), 52-58. I wish to thank the editors and publishers of these journals for permission to reprint.

CONTENTS

INTRODUCTION

Some Aspects of Conrad's Later Fiction

Some Aspects of Conrad's Later Fiction

The subject of this book is the morality of Conrad's novels after *Lord Jim.* When Conrad finished that largest of his early works, he felt he had written himself out. Yet in four years he published a novel almost as massive, namely, *Nostromo,* which Conrad was to consider his best. *Nostromo* resulted from a new kind of imaginative effort in which "research" outweighed experience. In *Lord Jim* Conrad used the biography of Sir James Brooke to give shape to his own Malaysian adventure, but in *Nostromo* he used numerous sources to detail the life of a continent he had visited only briefly. His lack of first-hand knowledge bothered him. He wrote to R.B. Cunninghame Graham, to whom he was indebted for references and personal reminiscences, that he felt a complete fraud.[1]

Scholarly studies have established Conrad's use of certain books about Latin America. This study can demonstrate, as a further illustration of the derivative nature of the later novels, Conrad's reliance on newspaper reports. In *Nostromo* Conrad identifies, I believe, the most important source of his political narrative when he refers to the *Times'* accounts and to the *Times'* "special correspondent."[2] These clues and the nature of the events lead one to the numerous dispatches published in the London *Times* during 1891 concerning the civil war in Chile. At least two of the dispatches were written by a "special correspondent," Maurice Hervey.[3]

1. G. Jean-Aubry, *Joseph Conrad: Life and Letters* (New York: Doubleday, 1927), I, 315, 337.

2. *The Works of Joseph Conrad* (London: William Heinemann, 1921), VII, 451. References to *Nostromo* in my text are to this edition.

3. Scholarly investigation of the sources of *Nostromo* has been summarized and added to by Norman Sherry in his *Conrad's Western World* (Cambridge: Cambridge University Press, 1971), pp. 147 -201. Sherry notes that Conrad borrowed the name *Moraga* from Maurice Hervey's *Dark Days in Chile;* but Sherry does nothing more with that text, nor does he examine the *Times* accounts.

In its departures from ordinary chronology and use of several narrators, Conrad's narrative method is not unlike the accidental method of successive news dispatches written with varying degrees of immediacy by several correspondents. One result in both instances is that events are somewhat obscured. For this reason I offer a brief synopsis as follows. In the novel Charles Gould, last representative of a prominent Costaguanan family of English descent, makes productive the San Tomé silver mine. Through its wealth Gould gains political influence in the Occidental Province, where the mine is located, and in the capital of the country, Sta. Marta. Eventually Gould and foreign commercial interests underwrite a revolution in favor of the reform president-dictator Ribiera, leader of a coalition of clergy and large landowners. Most of *Nostromo* concerns the later revolt of Ribiera's vain and Caesaristic Minister of War, General Montero. To oppose Montero, the Conservatives in Sulaco, capital of the Occidental Province, send south by steamer a force under General Barrios to the port of Cayta, his mission being to attack the flank of Montero's forces advancing on Sta. Marta. Another Conservative force under General Sotillo garrisons Esmeralda, only sixty miles south of Sulaco.

Ribiera, defeated at the Battle of Socorro, flees over the mountains to Sulaco. As soon as he arrives, he is set upon by a mob of Liberals and Nationals engaged in breaking the windows of the Intendencia. Rescued by Nostromo, Ribiera shelters with town officials in the O.S.N. offices by the harbor and eventually is taken off in a gig to the O.S.N. steamer *Minerva.* The mob attacks the jetty, the nearby Custom House and railway yards but is driven back into Sulaco and cleared from the principal streets by Nostromo's lightermen and European workmen. Earlier in the day a band of young aristocrats cleared the Calle de la Constitucion for the flight of women and children to two cargo boats in the harbor by firing on the mob in the Plaza from the windows of the Amarilla Club. Street fighting in Sulaco ends when the mob learns that Montero's brother will arrive soon with a Liberal force. The mob takes the road to Rincon to meet him. Another event of the "three days" of Sulaco is the defection of the Esmeralda garrison under Sotillo and its return to Sulaco.

Sometime later the Conservatives regain the town and the province according to a strategy devised by Decoud. Nostromo brings Barrios back from Cayta. As Barrios' steamer enters the harbor, Sotillo is stabbed by one of his own men. Montero's cavalry charges Barrios' force at the harbor gate, but new rifles procured for Barrios by Decoud turn the charge into a massacre. Simultaneously the San Tomé miners free Charles Gould, whom Pedrito Montero has ordered shot. Barrios pursues the Monterists, annihilating them at the Battle of Tonoro. The Conservatives declare the independence of the Occidental Republic.

Conrad's knowledge of Chilean politics came largely, no doubt, from Hervey's lengthy summaries in the *Times* of April 28 (p. 13) and May 19 (p. 8). Chilean Liberals, the dominant party, were opposed by the Conservatives. Like Hervey, Conrad calls the latter "Oligarchs" (e.g., p. 46); and he copies the composition of this "clerico-aristocratic" party in the alliance he invents between Father Corbelan, who, like his Chilean counterparts, seeks the restoration of church revenues, and Don José Avellanos, whose ties are with the provincial aristocrats. The alliance between Montero's Liberals and Gamacho's Nationals, further, recalls the anti-aristocratic and anti-foreign principles of the Chilean Liberals (see *Times*, January 21, p. 5) and an actual alliance of Liberals and Nationals in Chilean politics (see *Times*, May 28, p. 5). Charles Gould, the "king of Sulaco," must be modeled on a Chilean of English descent, Augustin Edwards, described in dispatches as the "richest man in Chile" and as the "nitrate king." Edwards' properties were confiscated (see *Times*, March 14, pp. 7, 15). Conrad rather obviously identifies Gould's prototype in this passage:

> So far this resembles the first start of the Atacama nitrate fields, where there was a financing house, a gentleman of the name of Edwards, and – a Government; or, rather, two Governments – two South American Governments. And you know what came of it. War came of it; devastating and prolonged war came of it, Mr. Gould. However, here we possess the advantage of having only one South American Government . . . (pp. 72 -73)

Conrad alludes to the war between Chile and Peru in 1880, whereby Chile annexed mining regions in the North. The passage points also to the civil war eleven years later ("one South American Government") as Conrad's further model for events.

Conrad alters the character and role of the Chilean President. Balmaceda was not, like Montero, a military man or a savage illiterate, nor did he gain power by a military coup. Conrad deviates from his source, also, in giving Costaguana a history of political turbulence. Before the civil war Chile was known to Europeans as the most stable republic in Latin America. The reason for these alterations is apparent. Neither the constitutional nature of the Chilean struggle nor the country's former stability lent itself to Conrad's ironic purpose, namely, to make the attainment of a democratic ideal depend upon the success of an ignoble materialism. Suggestions for Conrad's changes, however, exist in the Chilean materials. A startled reader in the *Times* of May 22 (p.13) reports that Balmaceda's character has two sides, one overlooked by the sympathetic *Times* correspondent, i.e., Maurice Hervey. Conrad may have conceived of Montero as Balmaceda's *alter ego.* A Peruvian admiral of the same name, who operated in the nitrate regions during the Peruvian-Chilean War, who almost became President of Peru, who had a rival Gamacho or Camacho, is also a possible model. Conrad could have read of him in accounts of the Peruvian war in the *Revue des Deux Mondes.*[4] The author, M.C. de Varigny, who also wrote for the *Revue* one of the later reports of the Chilean civil war,[5] emphasizes "*l'impéritie et l'arrogance*" of Montero.[6] Conrad, whose letters record that he read the *Revue,* appears to describe that source when he has Decoud contribute an article on the insurrection in Costaguana to a "famous Parisian Review" (p. 148; cf. p. 145). One of the *Deux Mondes* articles is by a Chilean.[7] A suggestion for the five years of probity in Costaguana before Montero's revolt exists in Chile's fifty-year history of responsible government following an era of "pronunciamentos." A simple condensation converts the historical pattern into the fictional.

The clearest journalistic account of the fighting in Chile is Maurice Hervey's letter in the *Times* of April 28. At the beginning, Congress, controlling the navy, attempted to capture Valparaiso; but the army, loyal to Balmaceda, successfully defended the city. Congress then turned its attention to the North. It bombarded and

4. Vols. 46 (1881), 350 -79; 48 (1881), 654 -79; 63 (1884), 406 -33.
5. Vol. 108 (1891), 406 -35.
6. Vol. 48 (1881), 675.
7. Vol. 105 (1891), 909 -21.

captured the nitrate port of Pisagua, forty miles north of Iquique, the capital of the province of Tarapacá. Pursuing government troops inland, forces from the fleet were routed at Zapiga (a town in *Nostromo* [p. 281] has the same name); and government troops immediately retook Pisagua. Pisagua soon fell, again to the insurgents, who this time gained a victory inland at San Francisco. The garrison of Iquique came to the aid of the government troops, leaving the provincial capital undefended; and consequently the fleet occupied Iquique. Colonel Robles, leader of the government troops, defeated the insurgents at Hauraz; a part of his force under Colonel Soto pursued the insurgents to Iquique and regained the town. The fleet then bombarded Iquique until the English Admiral Hotham arranged a truce and Soto's surrender. In the final action in the North the insurgents defeated Robles at Pozo al Monte.

Conrad begins his chain of events in the North rather than the South, relating their progress from that geographical perspective. He alters the location of cities; he reverses the direction of troop movements; he turns allegiances around; above all, he suppresses events at Valparaiso and Pisagua in favor of events at Iquique; but the broad outline of the Chilean war survives, nonetheless, in his narrative. The first action in *Nostromo* is at Cayta, a great distance to the south near the capital; the first action of the Chilean civil war was at Valparaiso. The next action in *Nostromo* (Sotillo's mission and defection) is at Esmeralda, near Sulaco; the second action of the civil war was at Pisagua, near Iquique. The third action in *Nostromo* is in Sulaco, capital of the Occidental Province. The third action of the civil war was in Iquique, capital of Tarapacá. The final action in *Nostromo* takes place at Rincon, inland from Sulaco; the final action of the first phase of the Chilean civil war took place at Pozo al Monte, inland from Iquique. The situation of the Sulaco Oligarchs at the end of *Nostromo* is like that of Congress after it secured the North – separated from the rest of the country, protected by mountains and a desert, gathering strength for a further assault.

Conrad, it appears from the parallels cited, drew upon analyses of political alignments and general summaries of the progress of the war; but most useful to him, one judges, were the detailed accounts of particular actions in the North. The fullest description of the battle for Iquique appears in the *Times* of February 27. On February 16 Congressional forces occupied the city. That night fighting broke out between a mob and a force made up of foreign residents and soldiers from the fleet. On February 18 further fighting occurred:

> . . . the revolutionary soldiers, who had captured Iquique, were surprised by a remnant of the Government troops and defeated. There was hard fighting on the Pampa all the morning, and fighting in the town until night. The Opposition troops held the Intendencia, while the fleet kept up a constant bombardment in order to protect the Intendencia and cover the landing of the marines. The business portion of Iquique was fired by incendiaries the same evening. Captain the Hon. Hedworth Lambton, of her Majesty's ship Warspite, landed under fire for the purpose of arranging an armistice and taking on board any of the remaining women and children who might desire to leave. An armistice was declared till noon on the 20th through the mediation of Rear-Admiral Hotham. On the afternoon of the 20th the Government troops were surrendered to the fleet by Colonel Soto, who had fought gallantly for the town, which was now in the possession of the revolutionists. No foreign residents were killed, and the women and children, residents of the town, are living mostly on the ships in the harbour, while the men are remaining in the city. (p. 3)

This dispatch supplies most of the action contained in the "three days of Sulaco"; indeed, one can count the three days more easily here than in *Nostromo.* Conrad uses the mob's attack on the Intendencia and the mob's firing of the business section as well as the counteraction of the European residents. The scene on the plain is like that Decoud describes from the Casa Viola. Conrad substitutes an armistice negotiated by the Chief Engineer, an Englishman, for the armistice arranged by the English Admiral Hotham. And Conrad alters the flight of the women and children from warships to cargo ships. Perhaps a detail of the Battle of Pisagua stuck in his memory. A long letter in the *Times* of March 31 (p. 11) reports that the women and children found refuge there on the P.S.N. steamer *Arica.* Ribiera's two-fold flight, which is not in the account above, may have been suggested by Hervey's descriptions (April 28) of insurrectionists fleeing over the mountains into Argentina and to ships in the harbor of Valparaiso.

The *Times* of May 30 prints the letter of an "insurgent leader"; its most interesting part concerns a further action at Iquique:

> On the 19th . . . Viceate Merino Jarpa, who commanded a small lot of

> 40 sailors that garrisoned the town, learned that a force of 300 to 400 men were coming down from the hills beyond Molle to attack and possess themselves of the city . . . . He resolved to fight and to resist, as he knew well how, with all the energy of his noble soul. He shut himself up in the Customhouse with his 40 sailors, and without water and food he fought 12 hours from 6 in the morning to 6 at night . . . . at 10 Merino had half his men wounded, surrounded by an enormous fire, devoured by thirst, and their ammunition used up. Two sailors offered to go on board to ask help and water, food, and ammunition. They passed through the thick of the firing, threw themselves into the water, and swam to the vessels of war, although wounded on the journey. At 12 o'clock a column composed of 50 sharpshooters of Taltal – all young men of good position – landed for the protection of Merino. In the midst of a terrible fire they landed, carrying barrels of water, food, and ammunition, and firing as they went. (p. 7)

In reading this passage, one feels very close to Conrad's inspiration for the two most gallant episodes of *Nostromo,* namely, Nostromo's heroic exploits and the attack on the mob made by the young aristocrats of the Amarilla Club. It required no great reach of imagination on Conrad's part, one judges, to extend Jarpa's activities from the Custom House to the area around the Intendencia and to relocate and reassign the actions of the sharpshooters of Taltal.

The more important accounts in the *Times* have been examined thus far, but the many minor dispatches also contain suggestive details. Señor Sanfuentes, Minister of the Interior and Balmaceda's nominee to succeed him as president (*Times,* January 19, p. 5) supplies the name and a part of the career of Conrad's provincial deputy Fuentes. The *Times* of February 10 (p. 5) reports that three hundred workmen of the Primitiva Nitrate Company liberated political prisoners, a probable inspiration for Charles Gould's rescue. The point is often made (e.g., *Times,* March 27, p. 3) that the insurgents enjoy the nitrate revenues even as the Sulaco revolutionists subsist on the wealth of the Gould concession. The garrison of Pisagua defected to the insurgent side (*Times,* March 31, p. 11), an event that may have suggested the reverse revolt of the Esmeralda garrison. And accounts of the landing of government troops at Patillos (*Times,* February 10, p. 5) as well as at Iquique (Hervey's account of April 28) may have helped Conrad to imagine the odysseys of Sotillo and Barrios.

In Pisagua on the eve of the government occupation (*Times*, March 31, p. 11) eighty people took refuge in the English consulate, a detail reminiscent of the two scenes Decoud describes in the Casa Gould (pp. 220ff.) just before the arrival of Pedrito Montero. Conrad seems to have taken from descriptions of the Battle of Pozo al Monte the looting of Anzani's store (p. 448): "Later on, scattered bodies having secured arms from the battle-field, attacked and looted the *officinas* and stores, as well as a large bodega, the property of Señor Uvillas" (*Times*, May 16, p. 7). Conrad's detail of fugitives in the Custom House of Sulaco is in this description of fighting in Iquique: government troops "fought their way into the streets and attacked the Custom-house, where . . . a number of English and American residents had taken refuge" (*Times*, July 7, p. 5). During 1891 the *Times* frequently reported the efforts of Baron Hirsch, a name Conrad gave to the wretched Jewish merchant from Esmeralda, to resettle Jews in South America (e.g., May 25, p. 5). These articles are usually in the vicinity of the *Times* articles on the war in Chile. Like Barrios the Chilean Congress won its victories by means of an improved rifle (*e.g.,Times*, August 31, p. 3; cf. *Nostromo*, p. 448). At the end of hostilities in Chile, the United States (*Times*, October 31, p. 3) was the first to "acknowledge the Junta"; but anti-American sentiment ran high and the U.S. warship *Mohican* was readied at Mare's Island. These details seem to have run together in Conrad's imagination so that in *Nostromo* the U.S. warship *Powhattan* is the first to salute the flag of the Occidental Republic (p. 458).

Points enumerated above lead to the conclusion that the essential details of Conrad's political narrative in *Nostromo* are in the *Times* accounts of the Chilean civil war, albeit Conrad frequently modified them in order to clarify visual images and to achieve a more concentrated narrative.

Reliance on subject matter outside his own experience, which the discussion above documents in some detail, is a first general characteristic of Conrad's novels after *Lord Jim*. The novels vary in this respect. *Nostromo* is the most dependent, seemingly; the *Secret Agent* next – Norman Sherry has shown how helpful Ford Madox Ford was in giving information about the English anarchists. *Under Western Eyes* uses much factual information concerning the activities of Russian nihilists and double agents in Russia and western Europe. But the degree of the derivative falls off in that novel and falls off more sharply in *Chance* and *Victory*, where Conrad revived his earliest subjects, the sea and the Far East.

II

A second general characteristic of Conrad's later novels is that they are about subjects of contemporary interest. The educated Englishman of the period learned from the London *Times* and other newspapers of innumerable South American revolutions like the one in *Nostromo* and was informed that English and American entrepreneural activities caused them more often than not. In spite of Czarist censorship, the educated Englishman read in surprising detail about the activities of terrorists in Moscow and Geneva, like those in *Under Western Eyes*, and of the Russian government's repressive measures. Newspapers carried day-by-day accounts of Scotland Yard's progress, or lack of it, in solving the Greenwich Park bombing, on which Conrad based the *Secret Agent*. They described the agitations of well-known feminists – Mrs. Pankhurst, for example – whom Mrs. Fyne in *Chance* resembles. Financial schemes and disasters like de Barral's in *Chance* and Morrison's in *Victory* were common fare.

The fact of Conrad's social concern is lost sight of sometimes because his novels are not propagandistic as are the writings of his contemporaries Ibsen and Shaw. Conrad wrote his novels strictly according to the principle of his chief master Turgeniev, who did not sermonize but in subtler ways drew out definite attitudes in his reader. Conrad invested contemporary subjects with ironies and complexities which in many instances have become inaccessible. One example is his characterization of anarchists and their associates in the *Secret Agent*. They exhibit traits described by the late nineteenth-century criminologist Cesare Lombroso, who classified anarchists as degenerates. The modern reader will miss that aspect unless he makes a scholarly effort; and even so, he will be perplexed by Karl Yundt's violent opposition to Lombroso's identification of criminals by physical characteristics and by Conrad's derisive treatment of Ossipon, who employs Lombroso's categories. When Winnie appeals to Ossipon after stabbing Verloc, Ossipon, measuring her features according to

Lombroso, calls her a degenerate "of a murdering type." Ossipon "invoked Lombroso, as an Italian peasant recommends himself to his favourite saint."[8] Between 1899 and publication of the *Secret Agent* in 1907, *Blackwood's Magazine* published four long editorials about anarchists.[9] In 1899 Conrad wrote that *Blackwood's* was always "open" to him and was "the only one for which I really care to work,"[10] a comment which seems to approve the magazine's editorial attitudes. Familiarity with both Lombroso and *Blackwood's* enables one to hear Conrad's voice with more of the percipience of Conrad's contemporaries. One stumbles less over the Yundt and Ossipon riddles and comes to recognize in these episodes a familiar kind of irony in a familiar kind of thematic context.

The *Blackwood's* articles rely obviously upon a classification of anarchists as criminal degenerates. Like the *Secret Agent,* they describe physical abnormalities according to Lombroso: the anarchist has a "low" forehead, a "flat-gazing" eye, a receding chin (CLXVII, 689). Stevie in the *Secret Agent* is the same physical type, and the articles' further descriptions of anarchists as "feeble" or "narrow" or "weak" brained (CLXVII, 688; CLXX, 560, 561) suggest how close to stereotype Conrad's mentally retarded character was. The articles name the anarchist's mental disability more particularly: it is "hysteria" or "instability" (CLXX, 564; CLXVII, 693); "perpetual excitement may still throw the feeble mind off its balance" (CLXX, 565). The language is semi-technical and conveys, in brief, analyses from abnormal psychology other than Lombroso's which Conrad also uses in delineating Stevie's character. In Henry Maudsley, the leading English criminal psychologist of the time, a class of deficients is characterized by "abnormal excitability and extreme mobility of mind and body, but without epilepsy, paralysis, or bodily deformity."[11]

Excitability in this example is an exaggerated impulsiveness, and Stevie's is called that in the *Secret Agent.* Like other abnormal

8. Reference is to the Heinemann edition, vol. VIII, 336.

9. Vols. 167 (1900), 688ff.; 168 (1900), 402ff.; 170 (1901), 560ff.; 180 (1906), 128ff.

10. *Life and Letters,* I, 277.

11. *Pathology of Mind* (London, Macmillan, 1895), p. 339.

psychologists of the time Maudsley is indebted to Herbert Spencer, who regards impulsiveness as a characteristic of savages and degenerates. Their limited, often a-social reflexes are not restrained by the reflexional interplay of a complex, highly evolved nervous system. Degenerates for Maudsley have atavistic nervous structures in this Spencerian sense. Torpor of mind is the pathological result in Maudsley of over-excitement (pp. 119, 282, 285); and Stevie after such a state sinks into "hebetude" (pp. 53 -54). Maudsley hypothesizes that a sound animality holds "in salutary check the vagaries of an over-sensitive and unstable nervous system" (p. 62); elsewhere, he notes the salutary effect of sexual life (p. 131). Conrad makes clear enough Stevie's deficiencies in these respects. Both Conrad and Maudsley distinguish attacks of "mania" from attacks of epilepsy. According to the novel, Stevie is excitable but does not have fits (p. 8). The type in Conrad and Maudsley is identical.

Maudsley lists "mobility of mind" in the passage quoted above. A passage in Lombroso's *Les Anarchistes* helps one to understand better what meaning such a phrase might have in Stevie's case. Stevie's emotions alternate within fixed limits: "the tenderness of his universal charity had two phases as indissolubly joined and connected as the reverse and obverse sides of a medal. The anguish of immoderate compassion was succeeded by the pain of an innocent but pitiless rage" (p. 190). In the latter mood Stevie commits act of violence. Lombroso describes the same phenomenon in the actions of the anarchist Vaillant: "Cette grande mobilité, cette instabilité qui est le propre des hysteriques, on la retrouve dans ses convictions, dans ses changements continuels de mêtier."[12] Lombroso puzzles over the alternating of compassion and ferocity in other anarchists and explains it in this way: "Cela est dû à ce que notre moral comme nos nerfs est sujet à la loi des contrastes: une fois la bonté epuisée, on incline vers la méchanceté; une fois la cruauté satisfaite, on incline vers la bonté, de même que la rétine fatiguée du rouge voit du vert et *vice versa*" (p. 149).

Descriptive phrases in the *Blackwood's* articles most often call Stevie to mind, but they suggest other Conradian characters. A depiction of anarchists as "having no chin, and very little (when

12. *Les Anarchistes* (Paris: E. Flammarion, 1896), p. 68.

not too much) forehead" (CLXX, 563) sketches incidentally the physical type assigned in the *Secret Agent* to the Professor (p. 68). The hydrocephalic criminal is common to *Blackwood's,* to Conrad, and also to Lombroso. The Professor is, in another *Blackwood's* phrase, one of the "degenerates who deal in explosives" (CLXXX, 129). The anarchist in *Blackwood's,* further, is one "whose pocket is generally filled by his friends, and who is often accompanied by a faithful mistress" (CLXVIII, 403). A small ellipsis converts the passage into a description of the erotic Ossipon whose mistresses supply him with money. The latter part of the description reminds one also of Karl Yundt and his ancient mistress; and the first part reminds one also of Michaelis who allows himself to be patronized by the great lady. "The ex-prisoner could moon about the shady lanes for days together in a delicious and humanitarian idleness" (p. 57). Even as Michaelis is a chief example of a pervasive indolence, the Professor is a chief example of that pervasive vanity, amounting in his case to egomania, that *Blackwood's* and Lombroso detect in anarchists. The Professor has "such an exalted conviction of his merits that it was extremely difficult for the world to treat him with justice" (p. 83). *Blackwood's* complains, further, that the literary anarchist disregards the "rules of grammar and good sense" (CLXVII, 693). Lombroso cites the argot of anarchist writing as a degenerate symptom. And in the *Secret Agent* Vladimir scoffs to Verloc, "Some of your revolutionary friends' effusions are written in a *charabia* every bit as incomprehensible as Chinese" (p. 27). Presumably he has most in mind Michaelis, who in prison has lost the ability to think or write connectedly. Old Yundt illustrates a more general verbal difficulty that Lombroso (p. 50) also notices in anarchists: Yundt's "enunciation would have been almost totally unintelligible to a stranger" (p. 46).

Members of the revolutionary circle are superficial, cruel, vain, indolent. Their humanitarianism is, in a *Blackwood's* phrase, a "sham sympathy, rooted in selfishness" (CLXVII, 688), words that recall Conrad's statement of purpose in his letters (II, 59): not to attack the philosophy of anarchism but to show that anarchists and nihilists are "shams." In another passage in *Blackwood's* one seems to measure how completely the journal articles contain Conrad's conception of character and character relationship: "But if we blame the poor half-witted creature who strikes the blow, what shall we say of the cowardly miscreant who arms the trembling hand? " (CLXVII, 692) One might believe that this passage suggested to Conrad Verloc's manipulation of Stevie in the novel's pivotal action. That relationship is not historical. The Frenchman Bourdin, author of the Greenwich bomb outrage, to all appearances was his own agent.

Aspects of character and theme in discussion thus far are found in both *Blackwood's* and Lombroso. One aspect, namely, the anarchist (in the person of Michaelis) as "mystic" and "fanatic" occurs in Lombroso but not in *Blackwood's.* In *Les Anarchistes* some anarchists, unusually sensitive to human misery, live saintly lives; in another age they would have been hermits. They desire to suffer (pp. 101 -55). Conrad's epithets paraphrase Lombroso and add a vein of irony. Michaelis' glance has the "character of seraphic trustfulness" (pp. 53 -54); he has "innocence" of heart and "simplicity" of mind (p. 119); he resembles "those saintly men whose personality is lost in the contemplation of their faith" (p. 119); he is the "hermit of visions in the desert of a penitentiary" (p. 54).

Some aspects of character and theme, products of the English environment, match descriptions in *Blackwood's* but have no equivalents in Lombroso. Conrad's treatment of the English police corresponds in two important ways to the *Blackwood's* treatment of them. In the latter the policeman, aware of the bomb in the anarchist's pocket, "hesitates to arrest him, because he deems it against the rules of the sport in which they are both engaged" (CLXXX, 131). It is a lopsided sport. The anarchist, *Blackwood's* complains, "does not understand the rules of Society's game" (CLXVII, 689). Inspector Heat's chance meeting with the Professor in a deserted London street might have been written as illustration. The Inspector, knowing that the Professor carries a bomb which he will detonate if arrested, is apprehensive; but what stays the Inspector is what stays the *Blackwood's* policeman: "If I were to lay my hands on you now I would be no better than yourself" (p. 105). The Professor answers, "Ah! The game! " The novel offers this further comment: "Not that the Chief Inspector was afraid of him; on the contrary, he meant to have him some day. But not yet: he meant to get hold of him in his own time, properly and effectively, according to the rules of the game" (pp. 137 -38). The *Blackwood's* observation that anarchists do not play according to society's rules is implicit in these passages from the *Secret Agent* and is repeated more fully in the contrasts Inspector Heat draws between anarchists and the more common kinds of criminals with whom he deals. The latter know and accept the rules of violation and penalty under a system of law. Anarchists refuse the penalty and seek not to violate but to destroy the law. It is obvious that Conrad has repeated this thematic vein in the *Secret Agent* from journalistic literature. Moreover, in depicting Heat's activities and character, Conrad has taken suggestions from contemporary accounts of the actual handling of the anarchist problem in England. *Blackwood's* lauds the surveillance of the police: "True, the Anarchists are allowed the freedom of Soho, but it is a freedom sternly tempered by the knowledge and control of Inspector Melville" (CLXX, 562; cf. CLXVIII, 404).

Blackwood's complains about the English tolerance of anarchists and disparages those who shelter them, citing as an example an editor, a "Don Quixote in sympathising with lost causes" (CLXXX, 130). Conrad has two equivalents, the first in rank being the great lady, who harbors Michaelis and almost believes his socialistic theories. Conrad's treatment is less harsh than *Blackwood's* because, one judges, he has found in his model a distaste for modern times which parallels his own: "She disliked the new element of plutocracy in the social compound, and industrialism as a method of human development appeared to her singularly repulsive in its mechanical and unfeeling character" (p. 124). Another equivalent is the Assistant Commissioner, an idealist who holds Inspector Heat rigidly to the rules of the game and a chivalrous aristocrat, who obliges the great lady by protecting Michaelis. The word "Quixotic" in his case is appropriate.

From the parallels above one must conclude that the novel conveys a widely accepted view of its subject, albeit with fresh ironies and unexpected intricacies and nuances of motive and character relationship. And the novelist's sympathies and emphaties leave their imprint. But none of these obscures the fact that the anarchist to Conrad is despicable. Conrad even uses *Blackwood's* (CLXXX, 128) extreme phrase "mad dog" (p. 137) to describe him. And in the fashion of the day Conrad classifies the anarchists according to Lombroso as criminal degenerates. The reader who appreciates fully that Conrad made of his subject what was expected will find old Yundt's attack on Lombroso insignificant. It is in character; and, moreover, it adds to the novel's acceptance of Lombroso's anthropology no more than a minor reservation, which believers in Lombroso were usually willing to add also. A 1907 *Blackwood's* editorial article on the criminal, which follows Lombroso, still derives some typical mild amusement from Lombroso's disciple, who is "eager to measure the facial angle and to pinch the bumps of the criminal" (CLXXXII, 845).

The passage concerning Ossipon quoted at the beginning of this article may seem to satirize a superstitious reverence of a false prophet. But that reading, no doubt, is influenced by hindsight. Comparison of the *Secret Agent* and the *Blackwood's* articles makes it most unlikely that Lombroso had that character for Conrad. The passage is like passages in *Blackwood's* in which anarchists are said to be well read in Herbert Spencer (CLXVIII, 403; CLXXX, 129). These have a transparent irony in that anarchists are attracted to theories according to which anthropologists such as Maudsley classify them as degenerates. And such, I believe, is Conrad's intention in making Ossipon a disciple of Lombroso. It is a familiar kind of irony in a familiar kind of thematic context.

III

A third aspect of Conrad's later fiction is a qualification of what has been said concerning the topical nature of his subjects. In Conrad's last three important novels, *Under Western Eyes, Chance,* and *Victory*, in an increasing degree, the topical is displaced in favor of a more abstract and philosophical subject. In the last two, moreover, Marlow reappears as narrator possessed of a greater wisdom and more assurance than he possessed in *Lord Jim.* Conrad's essay into derivative subject matter may have been desperate, but it was also salutary. It was a phase in which he acquired a further degree of detachment in a Coleridgean sense, developing away from personal preoccupations into final maturity as a moralist. In varying degrees the last three novels have the same subject, namely, asceticism. In *Under Western Eyes* warfare between nihilists and the Czarist government gives way to personal relationships between the ascetic Razumov and the Haldins; in *Chance* a bogus financial venture gives way to feminism, which gives way to the relationship between the confirmed ascetic Captain Anthony and Flora de Barral. In *Victory* the history of the Tropical Belt Coal

Company gives way to the painful relationship between Lena and the arch-ascetic Heyst. The total effect of these novels is that of Nietzsche's famed polemic against the ascetic ideal; and that, I believe, is no accident. Even as Conrad's earlier novels respond to and embody Kantian and Schopenhauerian themes which late nineteenth-century intellectuals found absorbing, so his last three novels embody Nietzschean themes absorbing to early twentieth-century thinkers. The purpose of this study is to show that that is the case, that at the end of Conrad's writing career the special development of his moral insight was Nietzschean.

PART ONE: CHAPTER ONE
Ford and Conrad

Ford and Conrad

Recognizing the philosophical contemporaneity of an author who avoided explicit themes is a matter of regaining an earlier percipience, as in the study of Conrad, *Blackwood's,* and Lombroso above – discerning allusive phrasing and the special meaning of images. Accordingly, the second part of this study will be given to analyses of Conrad's last three major novels. The first part will establish a context for such analysis by describing Nietzsche's influence on some other members of Conrad's circle, especially Galsworthy and Wells.

One turns to Conrad's intimates because it was their habit much more than Conrad's to make philosophical references in their formal writings. In his letters Conrad offered criticism of many texts by Galsworthy and Wells, thus creating a kind of dialogue about morality for which the texts themselves often supply a Nietzschean reference. From these exchanges inferences concerning the Nietzschean or non-Nietzschean quality of Conrad's thought can be drawn readily. This inferential pattern can be applied also in a smaller way to Ford Madox Ford. Because Conrad did not comment on Ford's writings usefully in letters, Conrad and Ford's discourse must be reconstructed in some other ways.

In a letter Conrad wrote to Ford in 1903, Conrad describes the result of their collaboration playfully as an "over-man" and "what Nietzsche's philosophy leads to" (I, 313). The remark implies that one or both authors had studied the German philosopher extensively. Most likely at this time Ford was the leader. The son of a German expatriate, he had absorbed German culture as a student, as his reminiscences in *When Blood Is Their Argument* record. Of a period in Bad Soden in 1891, he reports, "I was under the care of a quite atheist and strongly Nietzschean Lutheran clergyman, who was my tutor."[1] The tutor, when he preached, "went as far as he could in the direction of ruling out compassion

1. *When Blood Is Their Argument: An Analysis of Prussian Culture* (New York: Hodder and Stoughton, 1915), p. 157.

as a mainspring of human motive" (p. 157). Ford's reminiscence indicates that he was schooled at that time in cardinal Nietzschean doctrines. The effect was lasting, one judges, from the fact that Ford's 1915 analyses of German culture are Nietzschean; the Prussian, for example, is the "Kulturmensch as superman" (p. 309). Ford consulted Nietzschean authorities prominent in the first decade of the century, it is clear, from his references to Frau Foerster-Nietzsche (*e.g.*, p. 145) and to Oscar Levy, editor of the 1909 -13 edition of Nietzsche in English (p. 310).

Ford was a Catholic; Nietzsche was a natural enemy. Ford avows that the Nietzschean creed "was a thing that I have never been able to consider as a factor influencing my own inner life" (p. 129). Ford says often that Nietzsche's thinking is confused. But Ford understood and appreciated Nietzsche's insights and gives him his due. Indisputably Nietzsche was an imaginative "genius" (p. 129). Nietzsche's philosophy had an advantage over all other modern philosophies except Schopenhauer's in that Nietzsche "did really consider that the only function of philosophy was to be the guide to life" (p. 130). Ford's most interesting comment is contained in this passage, in which Ford describes the ideas in *Thus Spake Zarathustra* as "anti-altruist, anti-Christian, and anti-religious, and the clear and convincing statement of the case against altruism, Christianity, and revealed religion was, at the time this book was written, well worth reading even for the altruist, the Christian, or the religious minded in general" (p. 132). Ford was attracted to Nietzsche because Nietzsche attacked English Utilitarianism and altruism. Ford found the morality of its chief exponents, Herbert Spencer and George Eliot, "demoralising" and described it, further, as "an attempt to combine the Christian standard of manners with a materialistic standard of values" (p. 132). Ford here modifies but does not obliterate Nietzsche's most frequent rationale. Nietzsche believed that altruism was a disguised form of the will to power; altruistic moralists thus were victims of self-created illusions.

Ford's phrasing above calls to mind *Nostromo*, which treats the suborning of idealism by materialism – the Goulds' failure – with consummate irony. Ford's more general laudation of Nietzsche calls to mind *The Secret Agent,* which treats ironically the liberal sympathies of the great lady and the Assistant Commissioner, distinctly English forms of altruism. It appears that these novels' criticisms of the modern world are related to Nietzsche's. While Conrad was writing them, he was on the most intimate terms with Ford. Ford's disclosure of the nature of Nietzsche's appeal for him enables one to identify as "Nietzschean" another attitude Conrad shared with Ford, namely, their vehement opposition to Tolstoy, a point which will be considered in the chapter on Galsworthy. Finally, Ford's disclosure gives a Nietzschean coloring to Conrad and Ford's agreement, as reported by Douglas Goldring, with respect to the *English Review* that there should be "no paltering with noble ideals."[2] While Ford was editor and afterwards, that periodical was receptive to Nietzsche. In 1911 Horace B. Samuel, translator of the *Genealogy of Morals* in the 1909-13 collected edition in English, contributed an article on the *Genealogy* (X, 130-44). In 1913 the *Review* published the Brandes-Nietzsche correspondence (XIV, 417ff., 535ff.). In 1914 appeared a prominent review of H.L. Mencken's *The Philosophy of Friedrich Nietzsche,* in which it was reported, incidentally, that "Nietzsche's philosophy has made great progress of late among all classes and throughout the whole world" (XVI, 583). In the period in which Conrad was more interested in the *Review* than in any other journal, there was a steady vein of Nietzschean reference in a variety of articles.

In the impressions assembled here Ford seems to run ahead of Conrad in fidelity to Nietzschean principle. That is probably a wrong conclusion. Ford relates that although he profited from reading Nietzsche, he did not become an anti-Christian. In 1914 Conrad wrote this statement to Edward Garnett: "Great, improving, softening, compassionate it [Christianity] may be but it has lent itself with amazing facility to cruel distortion and is the only religion which, with its impossible standards, has brought an infinity of anguish to innumerable souls – on this earth."[3] Both the substance and the phrasing recall Nietzsche's analysis of spiritualized cruelty and his attack on the ascetic ideal. These points occur in a letter to a writer familiar with Nietzsche's philosophy, one who appreciated fully the originality of Nietzsche's insights

2. *South Lodge: Reminiscences of Violet Hunt, Ford Madox Ford and the English Review Circle* (London: Constable, 1943), p. 24.

3. Edward Garnett, ed., *Letters from Conrad 1895 to 1924* (London: Nonesuch, 1928), p. 265.

into religion. The following tribute occurs in an essay in Garnett's *Friday Nights:* "His [Nietzsche's] analysis of the Christian and ascetic ideal in life, though extremely one-sided, is, on its special lines, the most telling and brilliant psychological analysis yet made."[4]

4. *Friday Nights: Literary Criticism and Appreciations* (New York: Alfred A. Knopf, 1922), p. 8.

PART ONE: CHAPTER TWO

Wells and Conrad

Wells and Conrad

As is well known, Conrad met H.G. Wells accidentally in 1896 when Conrad sought out the anonymous author of an article on *Outcast of the Islands* in *The Saturday Review*. Conrad described their subsequent friendship in a letter to Wells dated September 6, 1898, as follows: "I have lived on terms of close intimacy with you, referring to you many a page of my work, scrutinizing many sentences by the light of your criticism."[1] Conrad here acknowledges in the most particular way Wells' tutelage. But critics have been reluctant to give the relationship its due, probably because of the unlike temperaments and disparate fiction of the two authors. I believe the relationship is significant for Conrad's late as well as for his early career and propose to examine the Conrad-Wells correspondence for whatever inferences can be drawn from it concerning Conrad's character as a moralist in *Under Western Eyes, Chance,* and *Victory*.

The most useful letters to that end are those in which Conrad reacts to the fiction and non-fiction Wells published in the period marked off by the correspondence, that is, until November 28, 1905. But others, personal in nature, give a sense of a continuing intimacy and establish a context. A selection from the personal letters follows. Five days after the letter quoted above, on September 11, 1898 (I, 249), Conrad wrote to Wells to say he intended to move to Pent Farm; the thought that he and Wells would then be neighbors made him jubilant. On November 17, 1898 (I, 254), Conrad wrote that he had called on Wells because of Wells' illness but found no one at home. At the end of November, 1898 (I, 256 -57), Conrad invited Mr. and Mrs. Wells to visit on Saturday because Mr. Pugh was coming. On December 23, 1898 (I, 263), Conrad wrote that he and his wife had called "yesterday" in vain. On a Thursday in either February or March, 1903 (I, 310 -11), Conrad observed that the last time he saw Wells, Wells' wife "looked delightfully well," and Conrad inquired

1. G. Jean-Aubry, *Joseph Conrad: Life & Letters* (London: William Heinemann, 1927), I, 248.

whether the Wells baby was better. On October 2, 1903 (I, 319), Conrad wrote apologetically, "I have neither appeared nor written because I have been gouty for the last week." A letter of November 30, 1903 (I, 321), in the first place, replies to Wells' apology: "Indeed I did not expect you in this awful weather, especially as Jessie reported you with a cold." In commenting on his own illness Conrad discloses incidentally his feeling that there were times when he did and did not welcome Wells' demanding conversation: "I was laid by the knee (this time) the day after we travelled up together. I did not feel particularly bright even then, or else I would have succumbed to your blandishments and stayed for dinner and a chat in town." Finally, the letter expresses thanks to Wells for good offices with the directness Conrad's acknowledgments of September 6, 1898, have: "It strikes me, my dear Wells, that, in your quiet, almost stealthy way, you are doing a lot for me; if it were not for you a lot of people would not know of my existence, anything palpable, and still less of my involved form of narrative. I will be delighted to meet Prof. York Powell." Powell, an Oxford historian and student of literature, was known for his acquaintance with illustrious Russian exiles in England. In a letter to Wells of February 7, 1904 (I, 326), Conrad describes another celebrity of the time, Sidney Colvin, as an "acquaintance which I owe to you, my dear fellow, in the long list of your good offices." In 1905 Colvin persuaded the Stage Society to produce a dramatization of Conrad's short story "To-morrow." One last letter of October 20, 1905 (II, 25), offers these domestic details: "Jesse begs you to explain to your wife that the only reason she has not yet put in an appearance is simply because she has no longer the pluck to drive herself or even to be driven by the lad."

The letters are spread thinly but continuously over the period. Conrad tended to use them, doubtless, inasmuch as he saw Wells frequently, for somewhat formal arrangements and courtesies. Nonetheless, the letters give a full impression of the familiarity of the two households and of the two men, of visits and casual encounters, of one family's interest in and concern about the other's domestic affairs. They give an impression also of a circle of literary acquaintances the two men shared, some of which were made because of their friendship.

The latter impression, perhaps, should be strengthened; that can be done by taking one example from the correspondence between Wells and Arnold Bennett. In a letter of December 8, 1897, Bennett confesses that he was given *Almayer's Folly* to review for Unwin but did not read it carefully and so undervalued Conrad until Wells intervened: "I owe you a good turn for pointing out Conrad to me."[2] Bennett continues with praise for *The Nigger of the Narcissus.* Conrad and Bennett became familiar to the point where criticism of Bennett's works bulks large in Conrad's correspondence. In a letter of March 10, 1902, Conrad discusses Bennett's *Man from the North* and concludes by wishing to hear Bennett's opinion of his book "*viva voce.* Come when you can spare a day. I won't be likely to have forgotten the book. We shall be back at home after next Monday. We can put you up and next day I could deliver you safe at H.G.'s palatial residence" (I, 303). The details suggest an ensuing literary discussion of some intensity by three of the most important novelists of the time, who knew themselves to be such. In this period at least they had a solicitude for each other, sometimes combined with asperity; on March 29, 1904, Wells wrote to Bennett, the "Conrads are under an upset hay cart as usual, and God knows what is to be done" (p. 107).

One other fact concerning the Conrad-Bennett and the Conrad-Wells associations can be added from another source. Violet Hunt, Ford Madox Ford's common-law wife, in *I Have This To Say* reports that Arnold Bennett was "handy and useful about the Review,"[3] meaning the *English Review*, which came out in 1908. Conrad discussed details of its publication with Ford constantly at the beginning so that Douglas Goldring says he was "in a sense Ford's co-editor."[4] It has sometimes been thought that Conrad and Wells quarrelled. Clearly they drifted apart. Conrad moved, and after 1905 their correspondence lapsed. But like Bennett, Wells was active in the *English Review* at Ford's and, presumably, Conrad's instigation. Wells observed to Bennett on

2. Harris Wilson, ed., *Arnold Bennett and H.G. Wells: A Record of a Personal and a Literary Friendship* (Urbana: University of Illinois Press, 1960), p. 38.

3. Violet Hunt, *I Have This To Say* (New York: Boni and Liverwright, 1926), p. 28.

4. Douglas Goldring, *South Lodge: Reminiscences of Violet Hunt, Ford Madox Ford and the English Review Circle* (London: Constable, 1943), p. 23.

October 29, 1908: "Hueffer has written to me about his *English Review*" (p. 153). The reason for Conrad and Ford's interest and regard, Ford gives clearly enough in a survey of contemporary fiction he wrote for the periodical.[5] He judges Bennett and Wells the two chief practitioners in the English tradition established by Dickens, Thackeray, and others. He writes, further, "The most intellectually influential figure in England at the present day is that of Mr. H.G. Wells" (p. 666).

From the Conrad-Wells correspondence and other evidence, it can be assumed that Conrad many times in private conversations crossed over the terrain Wells marked off for himself as a thinker and that when in his letters Conrad reacts in brief fashion to Wells' intellectual texts, Conrad's remarks may be allusive.

II

Estimating Conrad's stance as a moralist is a matter of identifying where Conrad agrees with Wells and where he disagrees. Conrad is most enthusiastic in remarks on Wells' fiction and most opposed in remarks on Wells' intellectual texts. Ford recalls that at the beginning of his collaboration with Conrad, Wells' *The Invisible Man,* "the end of the 'Sea Lady' and some phrases that that book contained, and two short stories called 'The Man Who Could Work Miracles' and 'Fear', made up at that date all the English writing that, acting as it were as a junta, we absolutely admired."[6] Conrad and Ford collaborated in order to become more highly conscious writers of fiction, and for that reason Ford's remarks probably have most to do with Well's techniques. But probably they reflect also admiration for Wells as a moralist. Conrad's own, more particular reactions go in that direction. In February-March, 1903, Conrad wrote to Wells about *Twelve Stories and a Dream,* and on October 20, 1905, about *Kipps.* Conrad wrote that "Filmer," the first short story in *Twelve Stories and a Dream,* along with *Love and Mr. Lewisham* (1900) and *The Wheels of Chance* (1896), convinces him that at bottom Wells is

5. *The English Review,* 3 (1909), 655ff.

6. *Joseph Conrad: A Personal Remembrance* (Boston: Little, Brown, 1925), p. 39.

"an uncompromising realist. There is a cold jocular ferocity about the handling of that mankind in which you believe" (I, 310).

Conrad had in mind the following examples. Filmer invents a flying machine; but when the time comes to fly it, he is overwhelmed by the press, the notables and an aristocratic lady attracted by his fame. Filmer commits suicide, thereby causing the crowd to believe the invention is a fraud. But later it is flown successfully. In the *Wheels of Chance* Hoopdriver, a draper's assistant, is a poor thing altogether; but this "Heir of the ages"[7] possesses an adventurous spirit. On a holiday bicycle trip he rescues a young lady from a sophisticated scoundrel, Bechamel, who is trying to seduce her. Hoopdriver would have fought like one "to whom Agincourt was a reality and drapery a dream" (p. 160). In *Love and Mr. Lewisham* the hero falters in his academic career because he makes an unsuitable marriage. The three have in common the fact that all suffer from inner disharmonies. Filmer's intelligence outstrips his manliness; sexual desire undermines Lewisham's intellectuality. And in Hoopdriver a romantic spirit combines incongruously with physical inferiority and social servility.

In this fiction man is at odds not only with himself but also with society. Wells represents how inhibiting to instinct and intellect the mean domestic routines and dispiriting working conditions of lower-middle-class life were. Hoopdriver is imprisoned in them, and Lewisham sinks into them. Wells collected details from his own background and adopted characteristically the point of view of an aspiring young man such as he once was. He added irony, which Conrad much admired. In commenting on *Twelve Stories and a Dream,* Conrad refers to Wells' success in this respect and his own failure (I, 310). Conrad, of course, is the master ironist; but his remarks call attention to the many surface ironies which convey social criticism in Well's fiction. Conrad singles out Wells' best talents and ignores his faults. One understands why he and Ford overrated Wells' fiction.

Relentless authenticity, the personal note, and irony are strong in *Kipps,* which Conrad writes about in a separate letter. *Kipps* is the story of a lad raised by his uncle, a shopkeeper, and

7. See *The Wheels of Chance: A Holiday Adventure* (New York: Macmillan, 1896), p. 186.

his wife. The novel contrasts Kipps' narrow existence in that household to the freedom of boyhood excursions into nature. Having inherited wealth, Kipps sojourns for a time in the genteel world, where he discovers the vacuity of the leisured class. He chucks it all finally to become a bookseller, adopting a life style inspired by the material simplicity of working-class life. Conrad writes about the "justness" and the "human" quality (II, 29) of *Kipps* and also about Wells' "justice" and "humane" attitude. Wells manages to be not only realistic but also compassionate, especially in the treatment of Kipps' relatives.

Discussion thus far sets forth sufficiently Wells' character as a realist and, perhaps, as an "uncompromising" realist. But the phrase "cold jocular ferocity" seems to mark out some very special ground on which Conrad and Wells were agreed as realists. What is cold, ferocious, and jocular in Wells' view of mankind is that it is an evolutionary view explicitly stated. No other novelist of the time risks outraging the average reader so freely in this respect as Wells does. It is a distinguishing characteristic. In the *Wheels of Chance* beneath Bechamel's civilized exterior lurks a "Palaeolithic creature of simple tastes and violent methods" (pp. 141-42); the description applies also by implication to Hoopdriver. In *Love and Mr. Lewisham* Lewisham's swindling father in law, Chafferey, whom Wells selects to utter the wisdom of the book, states the predicament of Lewisham and the race as follows: "Nature is full of cruel catastrophes, man is a physically degenerate ape, every appetite, every instinct, needs the curb; salvation is not in the nature of things but whatever salvation there may be is in the nature of man."[8]

Wells' stress on disharmonies in human nature suggests, further, that his evolutionary view is not simply Darwinian or Spencerian but derives also from the German Pessimists. That fact is abundantly clear in the whole of Wells' writing and is confirmed in the fiction by some general references to Nietzsche. Miss Walsingham, to whom Kipps is affianced for a time, is "that almost divine Over-human"; and her brother, who embezzles Kipps' fortune, is an "Overman" (p. 261). Chitterlow, who dragoons Kipps into underwriting the production of his fabulously

8. *Love and Mr. Lewisham: The Story of a Very Young Couple* (New York: Frederick A. Stokes, 1899), p. 275.

successful play and so makes Kipps' second fortune, calls a roll of contemporary realists: Zola (p. 116), Ibsen and Maeterlinck (p. 90), and "All this Overman stuff, Nietzsche and ali that" (p. 159). Later on, it is said of him, "he had been reading Nietzsche, and he thought that in all probability he was the Non-Moral Overman referred to by that writer" (p. 196).

Conrad praises in Wells' fiction a realism with a strong evolutionary and Nietzschean reference, and Conrad's phrasing is percipient. It seems logical to conclude that Conrad's own view of human nature was influenced by Nietzsche's.

II

Early in the century Wells wrote three prophetic discourses: *Anticipations* (1902), *Mankind in the Making* (1903), and *A Modern Utopia* (1905). The series established him as England's best-known thinker. Conrad read all three[10] and in his letters offers specific criticism of the last two – that is, if Conrad's references are clarified. In the Jean-Aubry collection the editor conjectures that a letter without an exact date (I, 328), belonging to the period February to May, 1904, refers to *A Modern Utopia*; but this cannot be. That date is not right. Also, the letter discusses the "New Republicans," who figure in the first two works but not in *Utopia*. Comments like Conrad's occur in letters Arnold Bennett and Ford Madox Ford wrote to Wells about the same time concerning *Mankind in the Making.*[11] In his letters Conrad does not comment directly on *Anticipations,* but his opposition can be assumed from his criticisms of *Mankind in the Making* and from his comparative judgment of *Utopia* – of the three works, the "nearest to my understanding and the most commanding to my assent."

Much of *Anticipations* is taken up with forecasts of a materialistic kind. Wells predicts accurately most of the aspects of modern transportation and modern technology. He predicts the

9. *Kipps: The Story of a Simple Soul* (London: Macmillan, 1905), p.68.
10. See *L & L,* II, 16.
11. Bennett criticizes *Mankind* in a letter to Wells dated October 8, 1903 (p. 99). Ford in 1903 in a letter without an exact date refers, as Conrad does, to the "New Republicans." See Richard M. Ludwig, ed., *Letters of Ford Madox Ford* (Princeton: Princeton University Press, 1965), pp. 18 -19.

diffusion of great cities and the growth of suburbs. No doubt Conrad found these "anticipations" curious, but they are not of a clearly ethical nature and can be passed over. The central point of *Anticipations* is that men of the future will propagate.[12] Wells is entirely serious, and the proposition has philosophical weight. It is the decision of the man of scientific education:

> He will be a father of several children, I think, because his scientific mental basis will incline him to see the whole of life as a struggle to survive; he will recognize that a childless, sterile life, however pleasant, is essentially failure and perversion, and he will conceive his honour involved in the possession of offspring. (p. 105)

In the light of *Love and Mr. Lewisham*, in which Wells records much of his father's and some of his own struggle to satisfy sexual desire in the poverty attending the intellectual and pedagogical life, the proposition seems a formulation of instinct. But in *Anticipations* the basis of the belief is held to be the thought of Malthus and Darwin. Wells allows no other sanction in matters of morality; he rules out an "obstinately anthropomorphic theology" (p. 282) – evolution has destroyed the doctrine of the Fall (p. 290) – and modern spiritualism (pp. 294-95, n.). In the fashion of the nineteenth century, however, he bestows a quasi-metaphysical Necessity upon the more scientific order he foresees; he likens it to a "huge secular movement in Nature" (pp. 246-47). Procreation, further, will be controlled. Wells foresees selective breeding to eliminate "all that is mean and ugly and bestial in the souls, bodies, or habits of men" (p. 298). Scientific rationality, moreover, will dispel sexual inhibition; "we invest the plain business of reproduction with a mystic religious quality far more unwholesome than a savage nakedness could possibly be" (p. 303).

The second cardinal point of *Anticipations* is the preeminence of the educated. Wells foresees accurately the demand for technicians in a technological society. Their importance is such that an hegemony of technocrats will supersede democracy (p. 213). Much of this is in the nature of self-interest and egoistic assertion, but the rational basis, again, is held to be Darwinian. Evolutionary theory has destroyed eighteenth-century liberalism,

12. See the chapter "Certain Social Reactions" in H.G. Wells, *Anticipations of the Reaction of Mechanical and Scientific Progress Upon Human Life and Thought* (London: Chapman & Hall, 1902), pp. 103ff.

Wells hypothesizes. All men and all nations are not equal (p. 289).

Mankind in the Making[13] covers some of the same ground. Again life is a "texture of births" (p. 11). Many of the chapters prescribe conditions for raising and educating the children of the future. Wells' conception of a world state narrows; he thinks now of blocks, especially an English-speaking federation into which he hopes to entice the United States by abolishing the English monarchy and aristocracy. Further, Wells attacks the "cant of Nature's trustworthiness," which is "partly a survival of the days of Rousseau and Sturm" (p. 77). The reason is clear enough. Rousseau was the chief of the eighteenth-century liberals who believed that men are equal. *A Modern Utopia*[14] supplies another reason. There Wells writes about the "sweetish, faintly nasty slops of Rousseauism" and argues against Rousseau that "order and justice do not come by Nature" (p. 171). Wells thinks of the future in terms of a controlled evolution and a society planned by a rational elite.

Like his fiction Wells' discourses contain references to Nietzsche and, also, to Schopenhauer. Wells' attitudes are ambivalent; and the references reveal rather accurately where German philosophy influenced his thought and where his thought diverged from it. Schopenhauer's doctrine of the will, rather than Nietsche's, most obviously shapes Wells' view of human nature; in *Mankind in the Making* Wells calls Schopenhauer the "most acute" of men (p. 15). Some of Wells' own summaries of evolutionary theory are corrupted by Schopenhauerian distaste, as in this example from *Anticipations,* in which the first part is Darwinian and the latter Schopenhauerian: "Man is the creature of a struggle for existence, incurably egoistic and aggressive" (p. 220). But because Schopenhauer denies the will, he becomes an antagonist. The New Republicans discern life's competitiveness; "being men of action, they will feel nothing of the glamour of misery that irresponsible and sexually vitiated shareholder, Schopenhauer, threw over this recognition" (p. 296). Clearly Wells holds that men of the future will propagate because Schopenhauer holds they should not and because Wells is aware of the vogue of Pessimistic philosophy among turn-of-the-century intellectuals. *Anticipations* contains a

13. (London: Chapman & Hall, 1903).
14. (London: Chapman & Hall, 1905).

reference also to Eduard von Hartmann, (p. 1, n.) who based his predication of the future on Schopenhauer's.

Wells' affirmation of life, since it emerges from a Nietzschean kind of rejection of Schopenhauer as "sexually vitiated," must owe much to Nietzsche. Nietzsche's influence is felt, also, in Wells' principle of the dominance of an elite. Such ideas were commonly regarded as Nietzschean early in the century, as Wells himself records in *The New Machiavelli* in describing the announcement of his conversion to Toryism in 1909: "I came to my point that the reality of human progress lay necessarily through the establishment of freedoms for the human best and a collective receptivity and understanding. There was a disgusted grunt from Dayton, 'Superman rubbish – Nietzsche. Shaw! Ugh! ' "[15] Nietzsche dreamed, of course, of the mastery of natural aristocrats, whereas Wells envisioned the dominance of technocrats in a man-made environment. Clearly Nietzsche's principles were attractive to Wells, but he turns away from Nietzsche decidedly when in *Mankind in the Making* he phrases his most important point in this way: "men will rise to be over-men" by selective breeding (p. 39). The statement rejects the Nietzschean doctrine of the emancipation of the will; one surmises that Wells had Nietzsche in mind when Wells wrote about "this cant of Nature's trustworthiness."

Conrad reacts to the outline of Wells' thought sketched above in two letters, one dated October 2, 1903, and a second, discussed earlier, which is dated indefinitely. The first letter breaks off – Conrad says he is not well – after this observation:

> Now I've lived with your book for some days, I like it more. What surprises me is to find you so strangely conservative at bottom. In the end I don't discover more points where we are in contact, I felt all these as I went along on my first reading. The divergences which arise from the dissimilar sides of our natures become more definite in the process of thinking as was to be expected, because one does not profoundly examine where one cordially agrees. Generally the impression of *soundness* is strong all through. (I, 319)

"Conservative," used here by one man of letters to another to describe a work read twice, must convey a literate meaning. It is logical to assume that Wells, in the philosophical references that have

15. *The English Review*, 6 (1910), 274-75.

been examined, laid the groundwork for Conrad's classification and that it can be read as follows: Wells' first allegiance is to the established tradition in English thought, the Darwinian-Spencerian, a position strongly challenged at the turn of the century by the radical propositions of German philosophy. Conrad's classification favoring the latter is like Ford's espousing Nietzsche as a critic of Spencer and Eliot. The parallel results from a considerable agreement between Conrad and Ford in matters of morality. There are differences. Wells was more susceptible, in ways which have been examined, to Pessimistic philosophy than Eliot or Spencer. And Conrad was much more susceptible than Ford. Conrad's affinity can be traced in the fuller criticisms of the second letter.

There Conrad offers two chief objections. The first is stated as follows: "The immediate practical danger of which I am thinking is lest it should suddenly be said, 'Ah yes, The New Republicans – these are the people who look at the world as a breeding place' " (I, 329). The criticism is tactful, a "practical" consideration; but it conveys Conrad's belief that Wells' description of the nature of life is inadequate. It can be assumed that, given a choice, Conrad would have chosen a romantic experience such as undergirds Pessimistic philosophy rather than the clinical experience which Wells postulates. Conrad once observed that Wells and Shaw were digging the graves of "individual temperaments."[16] Conrad's objection implies faintly, also, that some may look upon the world as a place where men do not breed – in Wells' terms, an implication of Schopenhauerian negativism. In a 1903 letter to Wells, Ford states the case more directly: "As to whether it ought to convert one to New Republicanism: well, if one accepts your dogma that the perpetuation of the race of men is *good* & all that militates against it evil . . . . But. . . ." (p. 18).

Conrad's most important criticism is this: "Generally the fault I find with you is that you do not take sufficient account of

16. In a letter to R.B. Cunninghame Graham, Conrad writes: "The stodgy sun of the future – our early Victorian future – lingers on the horizon, but all the same it will rise – it will indeed – to throw its sanitary light upon a dull world of perfected municipalities and W.C.'s *sans peur et sans reproche.* The grave of individual temperaments is being dug by G.B.S. and H.G.W. with hopeful industry. *Finita la commedia.!* Well, they may do much, but for the saving of the universe I put my faith in the power of folly." (*L & L*, II, 12)

human imbecility which is cunning and perfidious" (I, 329). If Conrad's intention is gauged again according to Wells' opposites, the sentence reasserts the indelibly mixed nature of man between Will and Idea which caused Schopenhauer and Hartmann to despair. "Cunning" and "perfidious" suggest the deviousness by which animal nature attains its ends in Nietzschean analysis, and "imbecility" in context conveys a Nietzschean judgment and a Nietzschean paradox. The phrasing is not inspired by *Mankind in the Making,* where Wells' analyses are uncomplex.

The letter contains two incidental criticisms with Nietzschean import. At the beginning Conrad chides Wells for addressing "a sort of select circle . . . leaving the rest of the world outside the pale"; Conrad complains of "an intellectual exclusiveness – and also an exclusiveness of feeling" (I, 328). Conrad must have in mind the fact that Wells banned artists and "inner lighters" from the society of the future. Conrad's complaint is a reminder of the temperamental and philosophical differences between the two writers and of Conrad's observation that the dissimilar sides of their natures became more evident in thought. But all of this is preamble. With mild irony Conrad urges Wells to impose his ideas on a larger audience: "No one's position is too absurd to be argued with. An enlightened egoism is as valid as an enlightened altruism – neither more nor less" (I, 329). Conrad's phrasing and choice of example prove that he was philosophically literate. "Egoism" and "altruism" are important and encompassing words. The leading philosophical question of the turn of the century was expressed ordinarily as an opposition between English altruism and Nietzschean egoism. Presumably Conrad is thinking of Wells when he uses "enlightened altruism"; but it may be necessary to explain how an ethic based on the survival of the fittest can be called altruistic. The two things are compatible in Spencerian thought, where man evolves away from impulsiveness and toward empathy, a more useful faculty in a civilized society. Wells' thought is Spencerian. In *A Modern Utopia* he foresees the struggle for existence shifting from the lower to the higher faculties so that "the ambitious and energetic imagination . . . may become the incentive and determining factor in survival" (p. 155). Conrad's willingness to draw upon Nietzsche and his indebtedness to both schools of thought are suggested by the even balance he constructs – one is as valid as the other, "neither more nor less.'

His detachment, again, is like that Ford displays in using Nietzsche as a counter to Spencer and Eliot.

The other incidental criticism follows immediately in Conrad's letter: "The principle of absolutism did not fail to maintain itself because there is anything absurd in absolutism, but because autocrats had made themselves unbearable through a sheer want of intelligence" (I, 329). The reference seems to be to Wells' abolishing monarchy and aristocracy. If as seems likely, Wells was inspired by Nietzsche to formulate a doctrine of the elite, Nietzsche would not have acknowledged them as such. Nietzsche excoriated scientists as modern equivalents of medieval ascetics. As has been suggested, Nietzsche's masters were not students like Wells but aristocrats. In the period Nietzsche is a chief spokesman for absolutism and a strident critic of democracy and, also, of anarchy, which is what Wells really forecasts. Conrad's defense of absolutism in the context established by Wells' references seems very Nietzschean.

Thus far Conrad's two letters on *Mankind in the Making* have been examined and various inferences made concerning Conrad's position as a moralist from his disagreements with Wells. This survey of the Conrad-Wells dialogue would not be complete without examining two other letters whose comments are more positive. The first letter confirms some points already established, and the second furnishes an insight into what solution Conrad would offer to the disharmony of human nature. The first letter is about *A Modern Utopia,* and the second seems to be about Wells' lecture entitled "The Discovery of the Future." In any case, that is Conrad's subject. Reacting to *A Modern Utopia,* Conrad talks about a more "accessible" Wells (II, 16); and Wells is more accessible. In the Preface to *Utopia* Wells acknowledges that in *Mankind in the Making* he was guilty of much "rash" writing. Instead of banishing his opposition summarily as before, Wells allows a complaining "botanist" to accompany his protagonist over the Utopian planet. The botanist is romantically attached to the world as it is; dogs and women excite his most "copious" feelings. The botanist loathes not leaving things to chance and so inspires the condemnation of Rousseauism mentioned already. The botanist's view of human nature – and probably of nature – is like Conrad's. Men are the "scars of the past," the botanist argues against his companion, who is "always talking as though you could kick the past to pieces" (p. 359). Some aspects of the book seem concessions to Conrad's earlier criticism. Wells disavows the "human stud farm" (p. 182), for example. The protagonist's double concedes at one point that in Utopia "all the physical and emotional instincts of man are too strong, and his natural instinct of restraint too weak, for him to live easily in the civilised State. Civilisation has developed far more rapidly than man has modified" (p. 292). The men of Utopia, further, are not homogeneous as in other utopias. Wells has decided that Darwinian theory requires variations and continuing evolution. He describes four types: the poietic, the kinetic, the dull, and the base (p. 265). The categorization recalls Conrad's belief in human "imbecility." Moreover, the poietic includes both artists and scientists. What Wells clings to amid these concessions is the "sane" order of the future and the realization of human potentialities. Each inferior being on earth, muddled by false beliefs and imperfect understanding, has his more rational counterpart in Utopia.

In the letter that remains, Conrad describes Wells' lecture as "splendid" but wishes that Wells had emphasized Conrad's view of

> the whole value of the future . . . consisting in what we do endure and shape *to-day* . . . . The future is of our own making – and (for me) the most striking characteristic of the century is just that development, that maturing of our consciousness which should open our eyes to that truth – or that illusion. Anything that would help our intelligences towards a clearer view of the consequences of our social action is of the very greatest value – and, as such, a guide. (I, 323)

Conrad's very abstract observations become more meaningful when considered within the framework of Wells' philosophical references. It seems obvious that the "maturing of our consciousness" is that modern consciousness of the process of evolution Wells talks about in *Mankind in the Making* (p. 21). For both authors its articulation was the "most striking characteristic of the [nineteenth] century." That consciousness enables us better to understand the effects on human nature of social attitudes and institutions, which have developed haphazardly and not always beneficially – "what we do endure and shape today." At last we have a guide to meaningful "social action" and mastery of the future. Anyone seeking to improve man's lot will think of altering the nature of man or of his milieu. Wells is drawn to the former and Conrad to the latter. Conrad's choice is consonant with his emphasis on human imperfection and is pessimistic, but the degree of pessimism falls short of Schopenhauer's or Hartmann's. In the acceptance of man's nature as it is and in the intention to accommodate social attitudes and institutions to it and so relieve it, Conrad's position in the context is Nietzschean.

III

Conrad's disagreement with Wells over the perfectability of man calls to mind at once three aspects of Conrad's later fiction with reference to which Conrad and Wells can be compared easily. One is Conrad's use of Rousseau in *Under Western Eyes*. During his stay in Geneva Razumov reaches the end of his rope. He has ingratiated himself with the conspirators, pretending to be a revolutionary, whereas he is a Czarist spy. He has involved himself with Nathalie Haldin, pretending to have been her brother's friend, whereas he caused his torture and execution. Razumov realizes these entaglements lead nowhere and that he cannot by any ordinary means extricate himself. He begins to write a record of events under a statue of Rousseau at a table on a small island in the lake. Conrad throws the detail into relief so that the reader must conclude it is a piece of symbolic shorthand. Its meaning is Wells' meaning in *Mankind in the Making* and *A Modern Utopia:* reliance on nature and chance. Razumov, the illegitimate son of a Russian prince, in a precarious social position, believes he can make his way by non-involvement and cool reason. He discovers that reasonable behavior in a repressive tyranny is an unattainable ideal and that his judicious intellect is too frail to withstand the psychological pressures tyranny exerts over the individual. After taking his place beside Rousseau, Razumov acts from feeling and instinct. Whatever Conrad's final opinion of Rousseau may have been, it is as though Conrad here offered to Wells a large example to demonstrate that Well's trust in sanity, or reason separated from instinct, is misplaced. The careers of the heroes of *Chance* and *Victory* make the same point.

Secondly, Conrad and Wells had a common knowledge of the contemporary criminologist Cesare Lombroso. In *Mankind in the Making* Wells attacks the theories of Lombroso's followers, calling them "mattoid scientists" (pp. 52ff.); and in *A Modern*

Utopia Wells derides those who believe that "people with a certain sort of ear commit crimes of violence" (p. 323). It may be that Conrad found this attitude of Wells to be part of the "exclusiveness" Conrad complains of, for in the *Secret Agent* Lombroso's theories enter very fully into the characterization.[17] Wells' truly scientific mind demolished Lombroso's quasi-science; Conrad's education did not prepare him to reject in the same way theories which all the world believed in. But probably there is this further point of difference. Lombroso classified the criminal as a degenerate, and his views were sometimes associated with Nietzsche's view of modern Europeans as degenerates. Conrad's stronger affinity for the corrosive view of man in Pessimistic philosophy – stronger than Wells' – no doubt explains Conrad's less critical reaction to Lombroso.

A third point of contact between Conrad and Wells is the Russian scientist Elie Metchnikov. On November 20, 1903, Conrad wrote to Wells, "Don't forget to send me Metchnikov's book. I am really curious to see that" (I, 321). Elie Metchnikov, Russian expatriate, winner of a Nobel prize for studies in immunology, was one of the most eminent of contemporary scientists. Ford recalls sitting next to him at a dinner in London where Metchnikov's conversation was about the large bowel, its absence in birds and its hazardous presence in man.[18] Metchnikov thought man retained his food too long, and the solution and the key to longevity was a diet of fermented milk.[19] He was one of the earliest yogurt enthusiasts. Wells in *A Modern Utopia* in a passage on the effects of tea and coffee on the innards writes, "Rather would I be Metchnikoffed at once and have a clean, good stomach of German silver" (p. 67). W.H. Hudson wrote to Edward Garnett on one occasion promising to return a copy of Metchnikov he had bor-

17. See Robert G. Jacobs, "Comrade Ossipon's Favorite Saint: Lombroso and Conrad," *Ninteenth-Century Fiction,* 23 (1968), 74 -84; and Chapter III, "The Criminal Psychology of *The Secret Agent*" in my book *Joseph Conrad: The Making of a Moralist* (Amsterdam: Editions Rodopi, 1972), pp. 117ff.

18. *Return to Yesterday* (New York: Horace Liveright, 1932), pp. 75ff.

19. See especially the chapter "Lactic Acid as Inhibiting Intestinal Putrefaction" in Elie Metchnikoff, *The Prolongation of Life,* trans. P. Chalmers Mitchell (New York: G.P. Putnam's Sons, 1910), pp. 161ff.

rowed – "a most repelling book."[20] These allusions, which can be matched in the case of other authors, suggest how knowledgeable the members of Conrad's circle were in the thought of the time.

Metchnikov appealed to them and to Conrad primarily, one judges, because of his interest in disharmonies caused by evolution. Metchnikov judged that the Pessimistic philosophies of Schopenhauer and Hartmann were reactions to physical disharmonies. Incidentally, Metchnikov regarded Rousseau as an opponent of science representing a return to religion and mysticism.[21] Because of disharmonies, the young are especially prone to pessimism; Metchnikov offered as counsel his own development from pessimism to love of life. The most striking passages in his writings are reports of cases of somnambulism in which individuals perform physical feats they are incapable of normally but which their ape-like ancestors could perform.[22] Metchnikov thus conveys in a novel and arresting way the existence of the unconscious, which was established as a principle in late nineteenth-century philosophy and psychology. Not Nietzsche or Schopenhauer or Hartmann nor English writers on evolution succeed in conveying a sense of multiple selves in the human being as vividly. Conrad's interest in his letter to Wells is strong. One relates that interest naturally to "The Secret Sharer" and some other stories like it – "The Shadow Line" and even "Typhoon," if its scenes in the hold are interpreted symbolically. The "Sharer" introduces into fiction the extraordinary psychological phenomenon of a divided psyche, apportioned between the Captain and the Sharer, the conscious and the unconscious, the latter being lawless, atavistic, and seductive. Again Conrad and Wells share an intellectual source; but it seems to have led Conrad to apprehend most vividly psychological

20. Edward Garnett, ed., *Letters from W.H. Hudson to Edward Garnett* (London: J.M. Dent, 1925), p. 61.

21. Conrad's letter to Wells probably refers to *Études sur la Nature Humaine* published in Paris earlier the same year. For Metchnikov's comments on Schopenhauer and Hartmann, see, *e.g.*, Chapter VIII (Paris: A. Maloine, 1908), pp. 216ff. For comments on Tolstoy and Rousseau, see Chapter IX, pp. 263ff.

22. See *The Prolongation of Life*, pp. 200ff.

perils to which Wells – even though he represents disharmonies – was insensitive.[23]

23. This chapter is based on the published correspondence of Conrad and Wells. Their unpublished correspondence at the University of Illinois has been described and used by Frederick Karl in "Conrad, Wells, and the Two Voices," PMLA, 88, No. 5 (1973), 1049 -65, an article I came across after presenting my manuscript to the press. I judge from Professor Karl's comments that the unpublished letters would not cause me to alter my conclusions.

PART ONE: CHAPTER THREE

Conrad to Galsworthy

Conrad to Galsworthy

Conrad's correspondence with Galsworthy is fuller than with any other of Conrad's correspondents; and with the one exception of the correspondence with Edward Garnett, with whom Conrad often discussed practical publishing matters, it extends over a longer period of time. It contains domestic details: the birth of Conrad's second son, the illness of his first, Conrad's own illnesses; one letter mentions a Christmas pudding sent by Jessie Conrad to Ada Galsworthy.[1] The two families were on the most familiar footing. Apart from family visits, the two authors met on numerous occasions. Ford Madox Ford in his *Return to Yesterday*[2] recalls that Garnett gave a lunch every Tuesday at the Mont Blanc restaurant in London, which Galsworthy attended and "Conrad and Hudson, when they were in town" (p. 397). Literary discussions in Conrad's letters to Galsworthy, which are the subject of this chapter, are only a part, albeit the most precise part, of a larger exchange of ideas; and, accordingly, one expects to find in the letters, as in the Conrad-Wells correspondence, allusions and *a priori* assumptions.

Conrad offered encouragement while Galsworthy was trying to get published and later wrote practical criticisms of Galsworthy's work. At the beginning Conrad's tone is rather patronizing as in his letter of January 16, 1898, on Galsworthy's *Jocelyn.* Conrad praises its "distinguished literary expression" and "fidelity to the surface of life." Answering the complaint of a critic or reader that the book lacks profundity, Conrad asserts Galsworthy has "absolutely touched bottom"; "A fairly prosperous man in the state of modern society is without depth, – but he is complicated, – just in the way you show him" (I, 223-24). The argument is in the nature of reassurance, but it has a special truth. More critics than Conrad would have encouraged Galsworthy to think that his chief talent and only necessary profundity lay in a beautiful rendering

1. See G. Jean-Aubry, *Joseph Conrad: Life and Letters* (London: William Heinemann, 1927), II, 36, 43-44, 51-54, 64, 94.

2. (New York: Horace Liveright, 1932).

of surfaces. In a letter of November 11, 1901, which is mostly about Galsworthy's *Man of Devon,* Conrad acknowledged the end of Galsworthy's apprenticeship: "You've gone now beyond the point where I could be of any use to you" (I, 302). Conrad offered encouragement when *The Island Pharisees* was rejected and advised Galsworthy to make a "concession to the need of popularity" (p. 305); but on the whole after 1901 the relationship was one of equals. Conrad and Galsworthy discussed with each other ideas they had for stories and novels; they sent each other manuscripts; Conrad wrote an "appreciation" of Galsworthy published in 1906 in *Outlook* and Galsworthy, an appreciation of Conrad published in 1908 in *Fortnightly.* A passage Conrad wrote in September, 1907, meant to soften Conrad's objections to *Joy,* finely conveys their later relationship: "Your activity having become as it were an ingredient of my mental life, I can judge with the intimacy of a kindred spirit and with the detachment of a separate individuality" (II, 58).

Only one episode disturbed the evenness of the later relationship, namely, Galsworthy's rewriting of *Shadows,* later entitled *Fraternity,* in the face of Conrad's almost violent objections to it. In July, 1908, Conrad, who had been arguing mildly with Galsworthy over the outline of the work, wrote, "Let us shake hands on that last chapter" of *Fraternity*, adding, "I am awaiting the promised ms." (II, 70-71). His immediate reactions to it he set down in two extremely long letters. And when Galsworthy replied, Conrad wrote a third and attached a postscript. Galsworthy took up the matter with Edward Garnett, who advised him not to change the ending of *Fraternity,*[3] as Conrad wished; and Galsworthy stood his ground. Conrad was preoccupied with Galsworthy's novel; Galsworthy's creative activity, indeed, had become an ingredient of his mental life. Conrad detected in the work a personal development he judged to be perilous; in his agitation his tone became pedagogical and his criticisms severe. Garnett read *Fraternity* as fiction and gave technical advice; Conrad read it as morality – "before all it is the book of a moralist" (II, 77) – and questioned its values. Accordingly, this part of the correspondence affords a very special opportunity for estimating Conrad's own

3. Edward Garnett, ed., *Letters from John Galsworthy 1900 -1932* (New York: Jonathan Cape, 1934), p. 13.

stance as a moralist in the period when he was writing *Chance* and *Under Western Eyes*. This chapter will consider the bases of Conrad's criticism and some philosophical and literary inferences that can be drawn from them.

The setting of *Fraternity* is the upper-middle-class, artistic and literary world of London. Its chief character is Hilary, a writer, who is married to Bianca, an artist. The two feel passion for each other but are overly civilized to the point where they cannot have sexual relations. Hilary is attracted to his wife's model, a girl from the country, whom his wife painted as a street walker. He buys clothes and finds decent lodgings for her and protects her from Hughs, a lower-class neer-do-well. In these characters Galsworthy works out his conception of "shadows." Hughs in his lust is what Hilary in his refinement is not; and a similar contrast exists between the physical model and the esthetic Bianca. At the end Hilary leaves Bianca. He intends to elope with the model but then, having gone to her room, he decides not to. The book contains numerous other characters, but it suffices to mention only Bianca's father, a retired professor who is writing a book on universal brotherhood, and Martin, Bianca's young relative who has practical schemes for making the world sanitary.

Conrad's letters have passages where the argument is made tortuous by Conrad's extravagant courtesy, and that is true even of some passages in the letters on *Fraternity*. But most are forthright – the following, for example, in which Conrad excoriates Hilary's character:

> I don't think, my dearest Jack, that in the glow of inspired composition (I am speaking with the utmost seriousness, I am under the charm of it) you have realized the harrowing atrocity of his conduct. You have refined and spiritualized that poor wretch into a remote resemblance to those lunatics – there are such – who try to cut off locks of women's hair in crowds. He is a degenerate who is completely satisfied with the last scene with the girl and therefore with incredible villainy and a total absence of moral sense will act as you make him act. The strain on the reader is tremendous and the whole thing borders on the intolerable. (II, 78)

By "villainy" Conrad meant Hilary's raising expectations in the model of a better life but, more importantly, sexual expectations he could not fulfill. Conrad goes beyond Galsworthy to describe him elsewhere as "impotent." In another paragraph Conrad takes up the explanation of Hilary's conduct Galsworthy seems to offer in the novel, namely, "class." That motive, Conrad asserts, is "incredible." In offering it, Galsworthy as a moralist "betrays the very truth of things." Strong class feeling would have "stopped him long before." Having rejected this motive, Conrad reasserts the one he gives in the passage above: it is a "pathological case of spiritual sensualism" (II, 78-79). Conrad instructs Galsworthy, further, that morbid psychology "is a perfectly legitimate subject for an artist's genius. But not for a moralist," by which he means the novelist as moralist may not offer pathological behavior as exemplary or ideal.

Conrad understood Galsworthy's intention toward Hilary well enough, but in an earlier passage he offers another interpretation of the meaning of the book to impress upon Galsworthy that Hilary's action was intolerable according to what he believed were his own and Galsworthy's common assumptions:

> It is impossible to read a book like that without asking one self, – what then? Are we to be sanitists or write checks, or are we to let casual girls embrace our knees till our, what shall I call it? – refined desire is completely satisfied, our humanity at the same time not being imperative enough and not elementary enough, I may say, to knock her mercifully on her head afterwards? Or is this, – I won't say tale, – is this vision of life and things, so admirably imagined, so felicitously and poetically presented, is it a mere declaration, not of the vanity of things (that would be a too optimistic view), but of the utter futility of existence? Pessimism can no further go. (II, 78)

Conrad expresses a bewilderment any reader will feel at not finding easily a thematic center in the novel amid its disparate parts. He names most of the parts: Martin's social work, charity toward the poor practiced by several privileged characters, and these in combination with Hilary's protection of the model expressed again as both inadequate desire and inadequate compassion and all of this as prelude. Finally Conrad offers the possibility that the work expresses the futility of human life, knowing that, like himself, Galsworthy does not find that an acceptable attitude in the novelist. These remarks are cogent even though the reader works hard at times to extract their meaning. "Vanity of things" is difficult. It suggests how Conrad (but not Galsworthy), aware that charity may be an expression of vanity, might have viewed the actions of Martin, Hilary, and the occasional philanthropists. To identify imperfection is to hope for improvement is to be optimistic about life.

Conrad's condemnation of the idealization of morbid psychology and his insistence that the novelist must have a faith, "however crude, distorted or extravagant" (II, 78), are precepts understandable in themselves within the context of the letters; but they possess broader meanings and are drawn from the larger context of Conrad and Galsworthy's common discussion and understanding of the philosophical and literary climate in which they lived. The first point is indirectly a characterization of the habit of Tolstoy, who by 1908 had become one of the world's foremost moralists. He was especially well known in England. Chertkov had founded there the Free Age Press which published first editions of Tolstoy. And there were other publishers of Tolstoy in English – Vizetelly, for example. In 1904 Leo Wiener's English edition of Tolstoy's collected works appeared. Tolstoy excited great interest among most of the writers with whom Conrad was intimate, their interest being helped along by the fact that Edward Garnett wrote criticism and Constance Garnett translated isolated works. In 1908 Ford Madox Ford asked her to translate a story for the *English Review,* preferably "not of a social propagandist kind."[4] Conrad knew that Galsworthy admired and feared that Galsworthy would emulate Tolstoy and says as much in the letters on *Fraternity.* Of the conflict between morality and art that may develop as a consequence in Galsworthy's work, Conrad writes:

> This may lead to a certain uncertainty of intention which may conceivably lead further to a vain harrowing of your feelingss, – tho' I don't believe that it will ever lead you into the gratuitous atrocity, of, say, *Ivan Illyith* or the monstrous stupidity of such a thing as the *Kreutzer Sonata,* for instance; where an obvious degenerate not worth looking at twice, totally unfitted not only for married life but for any sort of life, is presented as a sympathetic victim of some sort of sacred truth that is supposed to live within him." (II, 77)

And the last thing Conrad says about *Fraternity* in the postscript to his third letter on the subject contains this reference: "I still think that the idealizing for moral purposes of a man capable of

4. Richard M. Ludwig, ed., *Letters of Ford Madox Ford* (Princeton: Princeton University Press, 1965), p. 30.

such a spiritual crime is too Tolstoyan to be approved by a plain man without tears and sensibility like myself" (II, 86).

Conrad's comparisons are apt. "The Death of Ivan Ilych" and the "Kreutzer Sonata" are tales of men who, like Hilary, feel ill at ease in married life. In the first, Ilych is sickened by the usages of aristocratic Russian life and, more significantly for Conrad, by the sexual obligation of his marriage, and eventually dies. In his decline he is cared for by a strong young peasant, whose attentions assume for Ilych the utmost importance. To a mind used to exercising the will effectively upon the challenges of life, the consequences of this failure of will must be an "atrocity" and "gratuitous" in the sense that neither the author nor the character has sufficient self-knowledge to identify accurately the cause and the motive. In the "Kreutzer Sonata" another misfit more or less forces upon his wife a relationship with a music teacher, then imagines they are having an affair, kills his innocent wife, and is exonerated by the courts. This fiction also for Conrad was a vivid experience; he had a tender feeling for women. His developed psychological insight identified the nature of the husband's "stupidity" and left him appalled as a moralist at the way in which "monstrous" occurrences originate spontaneously in morbid states. Conrad's phrase "sympathetic victim of some sort of sacred truth that is supposed to live within him" conveys with considerable bias and some distortion Tolstoy's iteration in his later work of alternatives to the vanity, frivolity, emptiness, and injustice of fashionable life. It is difficult to recapture the moral climate at the beginning of the century. It was one in which Tolstoy recommended continence to married couples and his counsel was considered seriously.[5] He made sexual conduct important in ethical discussion. Conrad wrote to Garnett on February 23, 1914, what is obvious in the criticism above, Tolstoy's "anti-sensualism is suspect to me."[6]

Fraternity in the situation it creates of an aristocrat protecting a woman of the lower class, their relationship dissipating

5. See Tolstoy's "On the Relation Between the Sexes" in Leo Wiener, ed., *The Complete Works of Count Tolstóy* (London: J.M. Dent, 1904), XVIII, 439ff.

6. Edward Garnett, ed., *Letters from Conrad 1895 to 1924* (London: Nonesuch Press, 1928), p. 265.

eventually, may remind the reader also of Tolstoy's propagandistic novel *Resurrection.* Prince Nekhlyúdov in that work is susceptible to Conradian analysis. Like Tolstoy he attempts to withdraw from fashionable life, more specifically from an accomplished liason with the wife of a general and a prospective one with his aunt's friend Mariette. Attempting to make amends for having in his youth seduced Katyúsha and thereby caused her to become a prostitute and convict, he follows her into Siberia, offering marriage, which she declines. In that action Conrad surely would have classified him as a "pathological case of spritual sensualism." Galsworthy's attack upon class distinction, spread broadly through his work, derives, one judges, from Tolstoy's. In *The Island Pharisees* (1904) Antonia and her relatives cannot arrive at true sympathy for Ferrand by breaking through the attitudes of the upper middle class and come to recognize the anarchistic spirit in him as a threat. In his distaste for his own class and in his aborted relationship with Antonia, Shelton resembles Hilary. Conrad counseled Galsworthy not to let Shelton write his novels (II, 81 -82). In *The Freelands* (1915) an aristocrat, Tod Freeland, by choice becomes a crofter, working the land as Tolstoy did on his estate at Yasnaya Polyana. Finally, Galsworthy acknowledges his indebtedness to Tolstoy overtly if inadvertently in expressing dislike for D.H. Lawrence's treatment of sex. Galsworthy wrote to Garnett, "the men we swear by, Tolstoy, Turgenev, Tchekov, Maupassant, Flaubert, France . . . only use the body . . . to reveal the soul."[7]

The full meaning of Conrad's second precept above, that a writer must have faith, also derives from earlier discussion of the intellectual climate in which Conrad and Galsworthy lived. Galsworthy would have recognized it as a familiar stand against the morality of Schopenhauer and Hartmann. As seen before, Conrad entertains the possibility that in the ending of *Fraternity* Galsworthy meant to illustrate "the utter futility of existence" and exclaims, "Pessimism can no further go." For Hartmann, of course, the history of humanity is a developing conflict between the intellect and the will, to be resolved by universal annihilation when an equilibrium is reached and humanity can will not to be. Because of the conflict existence is absurd. Hilary, like Bianca and

7. *Letters from John Galsworthy*, p. 218.

some other characters, is an advanced type whose life can be thought to illustrate the conflict and its resolution in impotence. Moreover, his inadequacy thwarts the novel's most viable characters, Hughs and the model. That effect, no doubt, Conrad had in mind when he wrote, "Pessimism can no further go." Conrad sees in Galsworthy's tale a usual interpretation intellectuals at the turn of the century put upon Hartmann's theories. It is the interpretation Hardy made in *Jude the Obscure,* for example. Conrad's application of "pessimism" is precise, and only someone equally familiar with Pessimist thought would have understood him.

Conrad's references above to Tolstoy's works, moreover, in themselves suggest that Conrad's precept was formulated in opposition to Pessimistic philosophy. Tolstoy, a learned writer, makes many philosophical allusions, the most important of which are to Schopenhauer and Hartmann. Anyone familiar with Tolstoy had encountered Pessimistic philosophy at least at second hand. This is especially true of the "Kreutzer Sonata." Tolstoy struggled against Pessimistic conclusions, to which he was highly susceptible, finding a credo eventually in New Testament teachings. But in the "Kreutzer Sonata" susceptibility is strong and Pessimistic theory is embraced. The story contains this consideration of Hartmann: "carnal love, will also be destroyed, and the prophecy will be fulfilled, people will be united, the aim of humanity will be reached, and there will be no reason for it to exist." In the same place the story voices a modern skepticism: "Why should we live? If there is no aim, if life is given us for life's sake, there is no reason for living. And if it is so, then Schopenhauer and Hartmann, and all the Buddhists are quite right" (XVIII, 342).

Thus far a condemnation of the idealizations of morbid psychology and an insistence that the novelist must have a faith have been taken from Conrad's unfavorable criticism of *Fraternity* and given a contemporary reference. Conrad's stance as a moralist can be gauged also from more positive assertions which occur in his criticisms of *Fraternity*. Conrad directs against effeteness a common judgment and, thereby, the validity of ordinary mores: "There will be thousands who will feel a sort of uneasiness. It will to them dwarf the greatness of the book. I am talking of simple people. The other kind will no doubt discover varied interpretations, not one of which will be worth having" (II, 78). Of the same kind is Conrad's conclusion that "Hilary is not a sufficiently big and human figure" (II, 80). His nature is atypical and deviant; in actual life such a nature excites suspicion. Conrad adds that such pure spirits usually end in degradation; they sit down in the mud; "and he will sit snivelling too, in nine cases out of ten, because pure spirits have no use for simple human dignity, which is made up of good and evil faced openly, grasped with full knowledge" (II, 81). Conrad here describes his own ideal. The male character combats the evil of its own ambivalence and discovers its own good, that is to say, its identity in the strenuous exercise of fundamental instinct. In strengthening his own nature, in resisting its subversion, man attains dignity, another word for "assurance," another word for self-knowledge, which derives from accomplishment. An abundant experience of life separates Conrad's ideal from Hilary. It also separates Conrad, the ex-sea captain, as an author, from Galsworthy and gives to Conrad's fiction a different character. It determines that Conrad will embrace the optimism of common people and thereby separate himself as a moralist from Schopenhauer.

Another positive assertion in Conrad's letter on *Fraternity* is assertion of the authority of Turgeniev, which Conrad knew would weigh with Galsworthy. Galsworthy's admiration was common knowledge in the Conrad circle. In *Return to Yesterday* (p. 275) Ford refers to Galsworthy's "study" of the Russian novelist. When Garnett first met Galsworthy, the latter was writing *Villa Rubein* in imitation of Turgeniev.[8] In that light the following passage in Conrad's first letter is, beneath the courtesy, a mild reproach: "A book like this demands many auditions. It is anything but simple. Its suggestions are multiple (In which by the bye it does not resemble the work of Turgeniev, whom the idle paragraphist proclaims your chosen master) and run in and out of each other" (II, 77). In two other places Conrad appeals to Turgeniev's example. He compares Hilary with two "incapables," the "loquacious" Rudin in the novel by that name and the "nervous" Nejdanov in *Virgin Soil.* These two, however, are not, like Hilary, impotents; "that cannot be said in the absolute sense. They were both necessary – to their society and even to the state, at any rate to the community. They had at least some intention" (II, 80). Rudin is an aging student, an intellectual, whose one-sided development

8. See the same, p. 5.

makes him impractical. When Darya Mihailovna opposes his marriage to her daughter, he analyzes his position, finds it hopeless, and gives in. His attachment to Natalya has developed from his need to utter fine sentiment rather than from a powerful emotion of love. His failure is not one of morbid psychology. He has the "intention" and is not "impotent." He inspires a legitimate interest in the novelist as moralist; such lovers of truth – Rudin dies in 1848 on a barricade in Paris – have their usefulness to society.

Nezhdanov in *Virgin Soil* is the illegitimate son of a prince and a governess. He is educated and given an allowance by the prince and the prince's legitimate sons. Nezhdanov's past history is like Razumov's in *Under Western Eyes.* Nezhdanov goes as a tutor to a country house, where he falls in love with his patron's relative and ward, Marianna. They elope and live together as brother and sister for some weeks in two rooms in a factory. Nezhdanov suffers from a nervous sensibility and has a mixed nature – he is a plebeian and an aristocrat, a revolutionary and a poet – which makes him "incapable" of finding his way in life. Unwilling to have Marianna marry so ineffectual a creature as himself, he commits suicide.

Conrad's third reference is to Bazarov in Turgeniev's *Fathers and Children.* What would make Hilary better, Conrad says, is some "inconsistency" in Galsworthy's conception, "a thing that would not (that need not) be expressed at all but somehow *felt.* You know that sort of inconsistency that one is aware of in Bazarov, in the inflexible Bazarov, say, somewhere, a hint of firmness" (II, 86). This is one of Conrad's courteous statements. There is no inconsistency either in Turgeniev's conception or in his character. That conceit Conrad uses to suggest Galsworthy has made Hilary too perfect, whereas he cannot abide the fact that Galsworthy has made him too weak. Bazarov, the son of a retired army doctor and small landowner, is a scientist and a philosophical nihilist, a masculine, "firm" figure; still the reader senses an insecurity in his abrupt treatment of his indulgent parents, his extravagant animosity towards Arkady's worldly uncle, even in the fact that he contracts a fatal disease after he has commenced a love affair with the extraordinary Madame Odintsov. He is a subtly dimensioned character drawn from life, and that is Conrad's opinion also.

Conrad's sensitivity to Galsworthy's quality is extraordinarily fine. He senses accurately that Galsworthy is falling away from Turgeniev, who has been for both of them a supreme literary model and example of wholeness and simplicity, and is following after Tolstoy, whose character is complex and flawed. Galsworthy wrote to Garnett in 1902 re *Anna Karennina* that Tolstoy "touches a new and deeper degree of self-consciousness and therefore of analysis" (p. 36). These diverging estimates of Russian authors exist elsewhere in the Conrad circle. Ford held with Conrad, as did Garnett; and Hudson went the way of Galsworthy. In *Ancient Lights and Certain New Reflections* (1911) Ford called Turgeniev "the most gifted and the most technically perfect of all writers."[9] His opinion of authors in the following passage from the same work is also Conrad's and if taken as a statement of indebtedness, is the exact basis for Conrad's being classified in the pages of the *English Review* as an international novelist and a contrast to Wells:

> Yes, it is impossible to say whether Turgeniev or Flaubert were the greatest of all novelists. They lived and unfolded their unprecedented talents in the same years, in the same city, in the same circle, filled with the same high ideals and high enthusiasms. And this is a very striking proof of how high effort in the arts flourishes by the mere contagion of contact. (p. 185)

Ford's dislike of Tolstoy will be considered in detail later.[10] In 1917 Garnett published a book on Turgeniev. This passage from the Preface, written by Conrad, is both men's estimate of Turgeniev, Dostoievsky, and Tolstoy in that order: "All his creations, fortunate and unfortunate, oppressed and oppressors are human beings, not strange beasts in a menagerie or damned souls knocking themselves about in the stuffy darkness of mystical contradictions."[11] The book caused Hudson to write to Garnett defending Tolstoy.[12] Hudson's opinions had altered somewhat, it appears, from those he held in 1901 when he wrote to Garnett

9. *Ancient Lights and Certain New Reflections: Being the Memories of a Young Man* (London: Chapman and Hall, 1911), pp. 185 -86.

10. *Memories and Impressions: A Study in Atmospheres* (New York: Harper & Brothers, 1911), p. 320.

11. *Turgenev: A Study* (London: W. Collins Sons, 1917), p. viii.

12. See *Letters from W.H. Hudson to Edward Garnett* (London: J.M. Dent, 1925), pp. 178-80.

with reference to Turgeniev's *Lear of the Steppes* about "that true, that right, realism in which the Russians are so great and of which we are I suppose incapable" (p. 15).

The divided opinions of the relative merits of Tolstoy and Turgeniev among Conrad's friends are a natural result, in the first place, of a contention between the two Russians themselves. When Tolstoy in 1855 first went as an author to St. Petersburg, Turgeniev and others welcomed him; but Tolstoy felt uncomfortable among the literati and avoided them. Turgeniev continued to admire Tolstoy's talents but during his lifetime criticized what he called Tolstoy's "eccentricities." Conrad's attitude is in a very particular sense a continuation of Turgeniev's. Divided opinions are natural also in a period when Turgeniev's preeminence, as maintained by Conrad and Ford, was no longer accepted by many critics. In July, 1906, Galsworthy wrote to Garnett (p. 177) that he had been reading Maurice Baring's *Landmarks in Russian Literature.* Baring estimates that Dostoievsky but not Turgeniev is now judged the equal of Tolstoy.[13] Turgeniev's "vision was weak and narrow compared with that of Tolstoy, and his understanding was cold and shallow compared with that of Dostoievsky. His characters, beside those of Tolstoy, seem caricatures, and beside those of Dostoievsky they are conventional" (p. 99). Baring, incidentally, points to Frank Harris as the leading representative of the Turgeniev cult in England (p. 124). In his book on Shakespeare Harris declared Shakespeare could not have written a Bazarov.

In asserting to Galsworthy and others the authority of Turgeniev, Conrad recommended realism of a distinct kind. Turgeniev's literary art is the art of sparse and evocative detail; he presents scenes and images of great clarity. Maurice Baring expresses that quality of Russian literature when he says "common sense" and "matter of factness" are important in the treatment of nature; there is a dislike of "extravangance and exaggeration" (pp. 25 -32). In Conrad's estimation these qualities are the mark not of callowness but of wisdom; as a return to simplicity and the common outlook, they are a means whereby the novelist as moralist avoids Tolstoyan and other aberrations, to which his nature is prone.

Prince Kropotkin in his *Ideals and Realities in Russian*

13. (London: Methuen, 1910), p. 78.

Literature distinguished Russian realism from French naturalism on two counts: first, Russians could not "understand realism as a description only of the lowest aspects of life." The statement does an injustice to the French, but it is true Turgeniev offered a comprehensive study of society, and that is an important point to make also about Conrad. Secondly, Kropotkin, with Zola's example in mind,asserts, "For us, realism could not be limited to a mere anatomy of society . . . the realistic description had to be made subservient to an idealistic aim."[14] Edward Garnett, reviewing Kropotkin's book in the *Speaker* of April 1, 1906, makes a little clearer Kropotkin's meaning. Turgeniev's writings are not overtly propagandistic; Westerners are amazed sometimes to be told Russian novels aim at reform, "The fact being that the artist, great or small, had only to paint realistically the life around him to create a fresh wave of the strong contemporary liberal and humanitarian feeling in the mind of his contemporaries,"[15] this regarding Turgeniev's *A Sportsman's Sketches.* Kropotkin offers, further, this interesting analysis of Turgeniev's method: "When some human problem had thus taken possession of Turguéneff's mind, he evidently could not discuss it in terms of logic – this would have been the manner of the political writer – he conceived it in the shape of images and scenes" (pp. 90-91). The implication of the literary art for the novelist as moralist is that he thereby avoids the excesses of a specialized emotion. Paradoxically, the particularity of his interest advances the universality of his concern.

A final characteristic of the realism Conrad recommends is both a point of literary art and, again, a requirement of the novelist as moralist. Garnett expresses it in his study of Turgeniev, whose achievement it is "to bring to light the complexity of motives and instincts which is always lying hidden beneath the surface" (p. 94). It is a very special effect of point of view, the penetration of a massive intelligence undeflected by inadequate self-knowledge, and unclouded by the aberrations of morbid psychology. In this regard Turgeniev is like Bazarov, only larger; and that is one reason Conrad uses him in his letters on *Fraternity.* Ford offers the same example in commenting to Galsworthy on

14. (New York: Alfred A. Knopf, 1915), p. 86. [First pub. 1905.]

15. "Literature. Causerie of the Week. Russian Literature," *The Speaker,* 14 (1906), 411.

Villa Rubein in a letter written in October, 1900. Both Conrad and Ford feared the superficial, the sentimental, the conventional, and the false in Galsworthy's work. Ford advised him to have "enough contempt" (p. 12). Conrad varied the advice in his letters. With reference to the *Man of Devon,* Conrad wrote, "You are really most profound and attain the greatest art in handling the people you do not respect" (I, 301) and somewhat later, "The fact is you want more scepticism at the very foundation of your work. Scepticism, the tonic of minds, the tonic of life, the agent of truth, – the way of art and salvation . . . . As against your people you must preserve an attitude of perfect indifference, the part of creative power" (I, 301). In talking about *Strife* Conrad, meaning much the same thing, talked about "felicity of conception" (II, 58). And in his Appreciation of Galsworthy Conrad praised his "irony" and "compassion."[16]

A moralist who around 1908 took a stand agaist pessimism and degeneracy and the idealizations of morbid psychology and advocated a realism sensitive to the complexity of motives and animal instincts beneath the surface of human behavior might well be thought to have absorbed in one form or another much of Nietzsche's philosophy. It is an obvious conclusion, but it can and should be supported by some additional arguments. The most important is the agreement which exists between Conrad and Ford's opinions of Tolstoy and Galsworthy. Ford writes: "Today we have a comparative cleanliness . . . no longer are we little men forced to run up and down between the mighty legs of intolerable moralists like Ruskin or Carlyle or Tolstoy."[17] Presumably Ford reacts here as he does to those other "intolerable" nineteenth-century moralists Spencer and Eliot from a Nietzschean bias. Today's "comparative cleanliness" has resulted from the purifying fire of Nietzschean skepticism towards idealism and altruism. Both Conrad and Ford urge upon Galsworthy skeptical realism in a degree and of a kind that can only be called Nietzschean.

In the second place, around 1908 being for Nietzsche was almost a corollary of being against Tolstoy. R.A. Scott-James, a

16. See Walter F. Wright, ed., *Joseph Conrad on Fiction* (Lincoln: University of Nebraska Press, 1964), pp. 88 -94. *Cf. L & L,* II, 32 -33.

17. *Memories and Impressions,* p. 320.

newspaper critic Ford knew well,[18] wrote for the February 1909 issue of the *English Review* a criticism of G.K. Chesterton's *Orthodoxy* in which he assumes that Nietzsche and Tolstoy are natural opposites, the first being the chief exponent of extreme egotism and the second, of extreme self-sacrifice.[19] In his 1915 novel *The Freelands* Galsworthy pairs them instinctively in Nedda's reminiscence of Hampstead conversations. "On Sunday evenings at home people talked about a future existence, about Nietzsche, Tolstoy, Chinese pictures, post-impressionism."[20] Nietzsche and Tolstoy, further, knew they were each other's antagonists, and the reader of either one would have known that also. Nietzsche includes Tolstoy's "pity" in his attack on altruism.[21] Tolstoy's works contain various references to Nietzsche. This description of Russian prison conditions in *Resurrection* would have caught the attention of many English readers:

> Only by a special cultivation of vice, such as is carried on in these institutions, could a Russian be brought to that condition to which the vagabonds are brought, who have anticipated Nietzsche's doctrine and consider nothing forbidden, and who spread this doctrine, at first among the prisoners, and later among the people at large. (XXII, 79-80)

Thirdly, Garnett's first *Friday Nights* essay, which furnishes insights into attitudes toward Nietzsche in the Conrad circle, suggests that Conrad's recommending Turgeniev as a moralist to Galsworthy was tantamount to recommending Nietzsche. As noted before, Conrad holds up Turgeniev's character Bazarov as a model. Conrad's reference is to *Fraternity*, the degeneracy of whose central character might naturally call to mind Nietzsche's vehement strictures against the degeneracy of modern Europeans. Bazarov is a nihilist and his skepticism is proof against spiritualizations of the kind to which Hilary is susceptible. In the first essay of *Friday Nights* Garnett places Nietzsche and Bazarov in the same category: "He [Nietzsche] begins as an enthusiast, the ardent and generous champion of other men's goals, the follower of Wagner

18. Ford's letters reveal the relationship, *e.g.*, pp. 25-26.
19. 1 (December 1908-March 1909), 596.
20. (London: William Heinemann, 1915), p. 62.
21. See Oscar Levy, ed., *The Complete Works of Friedrich Nietzsche* (London: T.N. Foulis, 1910), XIII, 203.

and Schopenhauer – he enters into the blankly materialistic stage of Bazarovism, or negation – he emerges again with the positive creative values of his creed of life as the will-to-power."[22]

Finally, that there is Nietzschean meaning and reference in Conrad and Galsworthy's exchanges can be argued, at somewhat greater length, from Conrad's reactions to Galsworthy's *The Patrician,* whose conception, I believe, is indebted to both Nietzsche and Tolstoy. Garnett criticized the novel, claiming that Galsworthy did not know his subject well enough. Galsworthy, piqued, submitted to Garnett a list of one hundred thirty aristocrats whom he knew or had met.[23] However, character in *The Patrician*[24] seems not so much true to life as true to Nietzsche's conception. Lady Barbara can toy with the idea of falling in love with the lower-class writer Courtier, but eventually she will marry one of her own kind. Her most characteristic moment is a moment of decision when she looks at London from the tower of the Caradoc's town house with a "feeling of power" (p. 300). Her mother, Lady Valleys, recognizes in herself the "pleasurable sensation" combat always arouses (p. 136); like the other Caradocs she discovers that the theory of the survival of the fittest suits the powerful. Lady Valley's mother, Lady Casterley, the most active and dominating of them all, exercises her will in furthering the political career of her grandson Lord Miltoun. As an embodiment of the aristocratic principle in the twentieth century, she possesses an "alert vigour which those who have nothing socially to hope for are forced to develop, lest they should decay and be again obliged to hope" (p. 200).

Galsworthy wrote of *The Patrician* to Garnett that in it he was "passing from satire to another phase." He had attacked class distinctions in earlier novels, but now "I wanted to go for the spirit behind the general spirit of things" (p. 181). The "spirit" Galsworthy seizes upon is the conflict, as the novel phrases it, between "what the philosophers would call the will to Love, and the will to Power" (p. 294). The analytical intention and the final conception are clearly Nietzschean. "Will to Power" is as unmis-

22. *Friday Nights: Literary Criticisms and Appreciations* (New York: Alfred A. Knopf, 1922), p. 6.

23. See *Letters from John Galsworthy,* pp. 7 -8.

24. References are to *The Patrician* (London: William Heinemann, 1911).

takable as Lord Dennis' casual reference elsewhere to "Henley and Nietzsche and that sort" (p. 120). The will to Love is represented by Mrs. Noel, a symbolic presence like Irene's in *The Forsyte Saga.* She is what the Caradocs are not: gentleness, passivity, sympathy, feeling, the love of beauty; and she is herself beautiful. She seems formed from the estheticism of Nietzsche's enemy Schopenhauer and the love-of-enemies principle of Nietzsche's enemy Tolstoy and, in an important degree, out of Galsworthy's own sensibility. Miltoun's relationship with her, which the Caradocs oppose, will lose elections for him; at the symbolic level the relationship is the psychological seduction of aristocratic vigor by pity and the love of beauty. Ultimately Miltoun follows the example of his sister Lady Barbara and clings to power.

Another Nietzschean contrast exists in the characterization, namely, that between Courtier and the Caradocs. Lord Dennis supplies a label for it when he distinguishes between those who live for ideas and those who live for facts (p. 152). Courtier lives for the former and Lady Barbara chooses to live for the latter. Like Nietzsche's plebeians, Courtier, the politician, cultivates and employs ideas to undermine the power of aristocrats.

One carefully placed detail of setting, finally, conveys a very special sense of the Nietzschean quality of *The Patrician.* In the library of Lord Valleys' country house, very visible on the wall, is the death mask of an Apache chief. It is described as an epitome of will and, as such, is emblematic of all the Caradocs, even little Ann Shropton. But what is most fascinating is Courtier's observation that the chief has no "soul" (p. 124). In the novel's Nietzschean context, that remark conveys the Nietzschean analysis that slaves have or get souls but not their willful masters.

Conrad's reaction to this novel differs from his reaction to *Fraternity.* He praises Galsworthy's psychological penetration: "You have put insides into these people with a vengeance" (II, 126). As detailed above, the special truth of character in this novel is Nietzschean. In Conrad's remarks on Miltoun, further, Conrad's assumptions seem very specifically Nietzschean.

> As to Miltoun, he stands out from them in a very striking manner, a very consistently sombre figure. You don't give him a single little gleam. He is to my mind more sombre than Bazarow [sic] and almost a plebeian, with his temperamental asceticism, his non-conformist conscience and his passion more like that of a priest in love than anything I

> could compare it to. He is a strange bird to come out of that nest. However, I don't suppose you meant him to be typical. He is a bigger creation than the others, but I should not say a greater. He is alive right enough, but the reader (this reader) somehow feels that he is what he is, because you *will him* to be so. Whereas the others exist as tho' they had come to life on their own. (II, 127)

What is most interesting in this passage is the use of "plebeian." It is Conrad's label and is not at all suggested by the novel. Galsworthy likens Miltoun's temperament to that of his sixteenth-century ancestor, an ascetic cardinal; and, also, he endows Lady Casterley with a similar rigor. Superficially at least, this temperamental vein in the Caradocs is more aristocratic than its opposite. But Conrad obviously believes that Galsworthy will understand his meaning. And his meaning is out of the ordinary in still another way – "temperamental asceticism," that is to say, "passion more like that of a priest in love" and a "non-conformist conscience." The average reader would know that in English history non-conformists were common people predominantly, but what ordinary reader would have believed the English plebeian to be an ascetic? The definition depends upon Nietzsche's theory that slaves by Christian asceticism achieve superiority over Pagan masters. Galsworthy surely uses that same theory in his study of the Caradocs, who for the most part have the temperament of the masters but in their history have absorbed that subversive form of the will to power represented by the priests. For that reason Galsworthy's characterization of the group has its own subtlety and considerable interest. But it is characterization according to the elaboration of a theory. And that, I believe, is what Conrad means when he says that Miltoun "is what he is, because you *will him* to be so."

PART TWO: CHAPTER ONE

The Moral Discovery of *Under Western Eyes*

The Moral Discovery of *Under Western Eyes*

Conrad's last three novels, published between 1910 and 1914, are strongly Nietzschean although in quite different senses of the word. The coincidence which exists between their publishing dates and those of the English edition of Nietzsche's Works (1909 -13) offers one explanation; and enthusiastic discussion of Nietzsche at the time in *The English Review,* as has been suggested, is another. A third is the fact that many English psychologists react in their works, *e.g.,* James Sully's *Pessimism* (London 1877), to the German school of philosophers, Kant, Schopenhauer, Hartmann, and Nietzsche. English writers oppose their psychological tenets especially to German pessimism with which Nietzsche is loosely connected. But reference to Nietzsche is frequent nonetheless. Conrad's fiction before *Chance* now and then gives off a Nietzschean gleam; but not until that novel is there a parade of Nietzschean words and concepts. This pattern brings one to the conclusion that Conrad had some interest in Nietzsche all along but did not really discover him until the publication of his works in English. Conrad's development has great consistency. His early works react against extreme idealism and pessimism from the bias of a modified Utilitarianism. His last works embrace Nietzschean realism and seek to reconcile a modified "altruism" with Nietzschean "egoism."

The narrator of *Under Western Eyes* asserts that "moral discovery" should be the object of every tale.[1] Presumably he speaks for Conrad; and presumably, from the insistent note the assertion has, that purpose of the novelist was impressed upon Conrad with added meaningfulness late in his writing career. Critics have not really described in *Under Western Eyes* a discovery of the magnitude the comment suggests, although they have marked out useful lines of inquiry, especially the unusual aspect of diabolism. By examining that aspect and some others with an insight some-

1. *The Works of Joseph Conrad* (London: William Heinemann, 1921), X, 82. References to *Under Western Eyes* are to this edition.

Reprinted from *Criticism* 14, No. 1 (Winter, 1972) by permission of the Wayne State University Press.

what different from that found in other discussions, this chapter seeks to understand more fully what "discovery" Conrad intended.

Albert Guerard has judged that Razumov's confessions at the climax, "I was given up to evil" and "I was possessed," are "rhetorical."[2] For this reason it is probably necessary to reexamine, largely in the novel's own terms, what evidence of diabolism Conrad offers. A passage describing Razumov's interviews with Mikulin, interviews which result in Razumov's becoming a secret agent, provides a strong clue to Conrad's conception of his character: "To the morality of a Western reader an account of these meetings would wear perhaps the sinister character of old legendary tales where the Enemy of Mankind is represented holding subtly mendacious dialogues with some tempted soul. It is not my part to protest" (pp. 380-381). Conrad's devil is as old at least as the Renaissance.

Sophia Antonovna, summarizing Razumov's career, refers to the "devil work of his hate and pride" (p. 474). The phrase describes presumably a particular interrelationship of passions in which the demonic can dress itself in modern literature and meet a modern demand for psychological complexity and naturalistic treatment of character. Razumov exemplifies pride in the sense of wilfulness when he rather pointlessly insists on making his own escape from Russia. But more characteristically pride diffuses itself into feelings of contempt and, in a sinister fashion, into feelings of extreme hatred. Razumov's sense of superiority generates contempt in St. Petersburg and Geneva for both revolutionists and reactionaries. Razumov, a man of the people on his mother's side, finds in the Russian peasant brutishness, inertia, and incapacity for action. Utopists who espouse the peasants' cause arouse his indignation. They would, by acts of violence such as Haldin's murder of the Minister de S, violate the "secular logic of human development" (p. 117). After his unsuccessful visit to the drunken Ziemianitch, Razumov reflects upon the contrast between the "clear grasp" of his own intellect and Haldin's "luridly smoky lucubrations." He is close to betraying Haldin. "No! If I must suffer let me at least suffer for my convictions, not for a

2. *Conrad the Novelist* (Cambridge: Harvard University Press, 1958), pp. 240-241.

crime my reason – my cool superior reason – rejects" (p. 41). The bizarre figures of the revolutionary circle in Geneva enlarge his sense of superiority: Peter Ivanovitch, whose feminist doctrines are too easily traced to sexual passion; Madame de S, whose spiritualism causes Razumov to protest to Peter Ivanovitch, "I am a serious person" (p. 266); Sophia Antonovna, whose shrewd insights lead her only to the false conclusion that Ziemianitch betrayed Haldin. Razumov measures the conspirators and others also by an exaggerated fastidiousness. Madame de S, with her painted face and claw-like hands, is a "witch in Parisian clothes" (p. 268). Tekla appears to have been "fed on unwholesome scraps and leavings of plates" (p. 289). Nikita is a "creature, so grotesque as to set town dogs barking" (p. 333).

A sense of intellectual superiority characterizes Razumov's somewhat less vehement scorn for the figures of reaction. Fashionable persons Razumov sees in the street just before betraying Haldin are not capable of "an act of conscience" (p. 46). Prince K's reserve toward his natural son excites Razumov's "indulgent contempt" (p. 63). Razumov loathes General T's suspicion and cruelty; the fanatical, goggle-eyed head of the police does not understand "a reasonable adherence to the doctrine of absolutism" (p. 104). The novel implies Razumov's isolation as one cause of such feeling. Haldin writes to his sister of Razumov's "unstained, lofty, and solitary existence," a kind of existence, the narrator observes, which makes a "young man shy and an old man savage." In isolation Razumov, through the enlargement of scrupulosity, grows unfit for human companionship. Sophia Antonovna chides him for his fault: "One does not know what to think, Razumov. You must have bitten something bitter in your cradle" (p. 316). Nor can he recover simplicity at the end. The independence he maintains from both parties, the reflection of an extreme complexity, prevents his being tied to mankind by affection and loyalty. His condition is that of a half-repentant Faust, to whom the devil in the guise of Mikulin once granted a personal interview.

Razumov's cleverness, perhaps his most striking demonic trait, is another manifestation of contempt. At the beginning of the novel, Razumov is a rather dull fellow, but after he betrays Haldin, his character undergoes a remarkable change. He is as witty and audacious as the Satan of *Paradise Lost* as he introduces the facts of his interview with Prince K and General T as hypothetical examples into his debate with Haldin. Or as he speaks half-truths misleadingly to the simple charlatan Peter Ivanovitch:

> Then aloud, with a satanic enjoyment of the scorn prompting him to play with the greatness of the great man:
>
> "Ah, Peter Ivanovitch, if you only knew the force which drew – no, which *drove* me towards you! The irresistible force." (pp. 284 -285)

Or as he deceives Sophia Antonovna out of her own mouth concerning the facts of Ziemianitch's suicide and Haldin's betrayal.

Razumov's extreme hatred of individuals can only be called malign, and sometimes it is motivelessly malign. Guerard has objected that his temptation to "steal the soul" (p. 448) of Haldin's sister – such non-naturalistic phrasing is common – is not credible. But it is not the only instance in which Razumov displays a radically abnormal and sadistic emotion. One thinks of the savage anger which causes him to beat the insensible Ziemianitch and one thinks of his homicidal fury toward Haldin. Razumov calls the thrashing "a sign of intimate union, a pathetically severe necessity of brotherly love" (p. 41). The phrasing suggests that in Razumov's solitary existence a fermenting desire for human relationships has been deflected toward the passion of hate, that the beating of Ziemianitch is a pathetic and perverted act of affection. Consciously or sub-consciously Conrad injects a similar complexity into the impulse to strangle Haldin which Razumov controls with difficulty during their last interview. Earlier Prince K, in the house of General T, had pointed out an exquisite sculpture, *The Flight of Youth,* which inspired in Razumov a "sensation resembling the gnawing of hunger" (p. 52). When Haldin pauses before leaving Razumov's room, the earlier description is recalled: Haldin "might have posed for the statue of a daring youth listening to an inner voice" (p. 76).

If the reader accepts the reality of these two episodes, he experiences no real difficulty in accepting Razumov's temptation to steal the soul of Natalie Haldin. She is Haldin's surviving image, and in their first meeting Razumov transferred his anger against Haldin to her, Razumov stresses in his confession. But his emotion is transparently more complex than hate. It is desire. He "gloated" over the possibility; he "brooded" (p. 449) ardently upon it. His emotion is somewhat reminiscent of Captain Anthony's in *Chance.* The solitary Anthony is overwhelmed by the prospect of possessing Flora more completely than any other woman. Considering the extraordinary range of civilized and savage components of the mind which Conrad's novels explore, one surmises that Razumov's fixations are thought of as manifestations of a primitive instinct for absorptive destruction returning in a sinister and refined form. To read Razumov's character in this fashion is to distinguish between the chains of rationalization whereby Razumov justifies his actions – rationalizations which do not take into account his compulsions to flog and kill – and the psychological and theological nuances with which Conrad invests his motives.

A few more demonic traits can be uncovered in the character of Razumov: a hatred of good in human personality in Razumov's several fixations, a pleasurable malice in the conversion of good into evil when Razumov manipulates Kostya into the theft of his father's money. Although their crimes are greater, one thinks of the malign guile of Iago and the ambitious cunning of Macbeth. The list of Razumov's demonic traits is substantial enough to cause the reader to take him at his word when he declares that he was "given up to evil" and "was possessed". And from the list one concludes also that the novel's references to diabolism reflect a conscious effort to reproduce some traits of the possessed as they are recorded in earlier literature.[3]

But to what end and in what spirit does this twentieth-century novel revive the demonic? Not at all, one decides, for the purpose of rehabilitating an ancient dogma and not at all in a spirit of devotion. The novel's point of view is enormously complex and interesting. And some space must be given to picking out its components.

These are most evident in the following passage, a part of which was quoted earlier:

To the morality of a Western reader an account of these meetings

3. Claire Rosenfield in *Paradise of Snakes* (Chicago: University of Chicago Press, 1967), pp. 145 -52, counters Guerard by treating the demonic as unconscious psychic metaphor.

> would wear perhaps the sinister character of old legendary tales where the Enemy of Mankind is represented holding subtly mendacious dialogues with some tempted soul. It is not my part to protest. Let me but remark that the Evil One, with his single passion of satanic pride for the only motive, is yet, on a larger, modern view, allowed to be not quite so black as he used to be painted. (pp. 380-381)

The point of view in the passage is that of the English narrator, who assumes here one of two voices he has in the novel. Confronted by the enormities of the Russian mind, he professes and displays a naïveté; but at other times he speaks with percipience, that is, with the voice of Conrad. Critics have recognized sufficiently this inconsistency, which is yet not a defect, for the reader accepts the contraction and dilation of the point of view, almost unconsciously, as a narrative convention. The quoted passage is percipient; the point of view may be said to have three dimensions. It conveys faintly the Russian attitude which finds the scene customary; it conveys an enlightened Western attitude which finds the scene unreal and falls back skeptically upon an archaic conception; and it conveys the author's encompassing attitude which points out playfully to the Western mind that the "larger modern view" passes beyond skepticism to feel, by implication, sympathy, familiarity, and kinship with the devil, specifically on the point of satanic pride.

Clearly this statement, above all others, expresses Conrad's attitude, but the statement is a puzzle. It does not convey finally an expected enlightened and naturalistic view of its subject. In exceeding the narrator's intellectual horizon, the passage exceeds contemporary English thought and the English philosophical tradition. English thought recommended altruism and disapproved of pride, another name for egoism. English rationalism might deny or discard the devil or, as in the quoted passage, relegate him to old, legendary tales; it did not seek to rehabilitate him, that is, to change his color. One assumes the metaphor makes good and exact sense. The passage causes one to search for an alternative to English Utilitarianism at the time Conrad was writing *Chance* and *Under Western Eyes.* Publication in English after a long delay of the works of Friedrich Nietzsche was, perhaps, the chief intellectual event of the period.[4] I wish to suggest that the "larger modern

4. As indicated before, there can be no doubt that Conrad knew Nie-

view" is essentially Nietzsche's view of the demonic, a view expressed by one of Nietzsche's most familiar titles, *Beyond Good and Evil.* In that work after stating his characteristic doctrine that our entire instinctive life is a "development and ramification of the Will to Power," Nietzsche manufactures this rhetorical question: "Does not that mean in popular language: God is disproved but not the devil? " He answers, "On the contrary" (XII, 52).[5] It is a way of recommending egoism. If God as good is disproved, the devil as evil is disproved. Nietzsche's upending of traditional values makes the devil as the patron of pride almost attractive; at least he is "not quite so black as he used to be painted." No other system of thought in the early twentieth century rehabilitates the demonic exactly in this way, that is, not by apologetics but by particular psychological analysis. There is no other answer to the riddle of the novel's more or less exasperating point of view, which has caused some critical confusion. The most important opinion at present is that for purposes of characterization, Conrad took from Dostoevsky and Tolstoy a supernaturalism he did not believe in or that Conrad's effort is to de-mythologize.[6]

The subject of point of view should not be left until the likelihood of Conrad's having adopted a Nietzschean attitude toward his subject has been reinforced. The narrator, the English teacher of language, in his most characteristic utterance professes not to understand the Russian temperament. He gropes for a way of describing it and produces, at last, the word *cynicism* – with a flourish (p. 82) so that the reader decides that the word conveys special meaning. And, indeed, it is not immediately understood with reference to characters and events. Conrad absorbed attitudes in his source materials. *Nostromo* echoes the anti-materialism of

tzsche. Conrad refers to him in his letters. See G. Jean-Aubry, *Joseph Conrad: Life and Letters* (New York: Doubleday, 1927), I, 313. Eloise Knapp Hay in *The Political Novels of Joseph Conrad* (Chicago: University of Chicago Press, 1963), p. 20, speculates that Conrad may have had "an interest in the historical works of Hegel and Nietzsche." Her discussion of *Under Western Eyes* does not mention Nietzsche.

5. References are to *The Complete Works of Friedrich Nietzsche* (London: T.N. Foulis, 1909 -13).

6. See, for example, Avrom Fleishman, *Conrad's Politics* (Baltimore: Johns Hopkins Press, 1967), pp. 225 -26.

some contemporary journalism devoted to entrepreneural enterprises;[7] the widespread journalistic classification of anarchists as degenerates[8] lodges itself in *The Secret Agent.* But the frequent newspaper dispatches and the numerous journal articles on Russia dating from the 1890's and the early 1900's do not use the word *cynical,* even though they describe the Russian people as "brutish,"[9] as does *Under Western Eyes. Cynicism* is a word with special meaning drawn not from popular writing or popular prejudices but from philosophical analysis. It is Conrad's word and not the narrator's, who, within his stated limitations, is incapable of so sophisticated a choice. Nietzsche I believe, does not anywhere call the Russian people "cynical"; but he uses the word in a relevant way in the following statement concerning the oppression of the slaves: "Cynicism is the only form in which base souls approach what is called honesty" (XII, 39). As in *Under Western Eyes, cynicism* in Nietzsche is nearly synonymous with *mysticism,* a renunciation of the world as it is and the seeking of very special compensations. When Oscar Levy wrote his commentary on Nietzsche's works for the last volume of the English edition, published in 1913, he summarized the political situation in the countries belonging to the Holy Alliance, which included Russia, in Nietzschean terms – forthrightly in terms of princely masters and slave populations (XVIII, ixff.). I believe that in *Under Western Eyes* Conrad did the same, applying Nietzsche's terms "cynicism" and "mysticism" naturally to the Russian people. In the Author's Note, Conrad says he did not possess "special experience" but "general knowledge" of the "moral and emotional reactions of the Russian temperament to the pressure of tyrannical lawlessness, which, in general human terms, could be reduced to the formula of senseless desperation provoked by senseless tyranny" (p. xi). In *The Secret Agent,* his last novel before *Under Western Eyes,* Conrad relied on contemporary psychological studies of anarchists,[10] of whom also he did not possess "special experience."

7. See, for example, *Blackwood's Edinburgh Magazine* CLXXI (1902), 840ff.

8. See, for example, the same, CLXXX (1906), 129.

9. See, for example, *The New Review,* V (1891), 45ff.

10. See Robert G. Jacobs, "Comrade Ossipon's Favorite Saint: Lombroso and Conrad," *Nineteenth-Century Fiction,* XXIII (1968), 74 -84; and Chapter III of my book *Joseph Conrad: The Making of a Moralist* (Amsterdam: Editions Rodopi, 1972), pp. 117ff.

One surmises that Conrad saw that Nietzschean psychology could be applied more readily than any other contemporary moral psychology to the situation in Eastern Europe. And one becomes convinced when he comes upon the Nietzschean word *senseless* used in a Nietzschean way in Conrad's statement of his intention quoted above. In *The Genealogy of Morals* Nietzsche writes of masters and slaves and of the craving for cruelty: "What really raises one's indignation against suffering is not suffering intrinsically, but the senselessness of suffering" (XIII, 77).

One other word is as important as *cynical* and *mystical* in the novel's analysis of the Russian temperament: Haldin's word, *resignation.* That is what is "divine in the Russian soul" (p. 26). Other characters, for example, Razumov, considering the drunken Ziemianitch, refer less flatteringly to the inertness and incapacity for action in the Russian masses. Haldin's "divine" poses a riddle. The complete set of terms would scarcely occur to an independent political analyst or novelist; and, again, it does not occur in journalistic accounts. The *cynical,* the *mystical,* and the *resigned* do not easily coalesce in human personality or in a national character, unless *resigned* means the inertia and sense of guilt in an oppressed people and *cynical* and *mystical* name that people's consequent dismay at the conditions of life and its Nietzschean recourse to a spiritualized cruelty, its creation of a "soul," and its transvaluation of values. In that context *resignation* becomes ambiguous and may be called "divine." This discussion illustrates the way in which a Nietzschean reading gives coherence to Conrad's view of his subject, whereas without it separate pronouncements are often baffling.

Nietzsche's genealogy of morals pictures in the early history of mankind the masters inflicting punishment upon slave populations in the interest of inculcating fixed ideas. Nietzsche terms the process "sinister mnemonics" because any curb to primitive impulse causes a painful complexity. "Alas for this mad melancholy beast man! What phantasies invade it, what paroxysms of perversity, hysterical senselessness, and *mental bestiality* break out immediately, at the very slightest check on its being the beast of action" (XIII, 113). In this condition man becomes interesting; by internalizing primordial instincts he creates a soul; through conflict between impulse and restraint man fashions a conscience and becomes evil. The animal ego turns against itself. In this "priestly" period of history religion ministers to the tamed animal by prescribing cautious gratifications of the will to power. Natural cruelty, which finds no natural outlet under oppression, is "deified," for example, in the festive spectacle of the *auto-da-fé* (XIII, 74). A chief result in base men of the oppression of the blonde beasts is "enervation" (XIII, 37). Like Nietzsche Conrad distinguishes the Russians from Western Europeans in that the Russians remain in a state of "resignation." Nietzsche interests himself, however, not in the slaves' passivity but in their revolt. They force subtly upon the masters a "transvaluation of values." All that is "manly, conquering, and imperious" and relishes life as it is is pulled down eventually; and the mystical and the ascetic are valued in their place.

Conrad constructs the character of Razumov according to these insights; Razumov becomes "evil" in the sense that Nietzsche calls Russians "evil" in Maxim 22 in *The Twilight of the Idols* (XVI, 4). Razumov points out that in autocratic Russia he is at the mercy of lawless forces. Conrad uses the word *resentful* to describe him even as Nietzsche uses it characteristically to describe the oppressed. Razumov's passive mental state after he has betrayed Haldin and, more importantly, after he has drawn to himself the attention of General T and savored his cruelty, exemplifies the enervation of the victims of the blonde beasts. "I am his prey – his helpless prey," (p. 59) Razumov declares in a distinctly Nietzschean mode of expression. Razumov's hallucinations recall Nietzsche's remarks (XIII, 171) on those mental disturbances that the methods of religion (starving the physical) beget in combatting physiological depression among the slaves. In part because of his designs upon Nathalie Haldin and also to escape isolation and inertia, Razumov agrees to spy upon the revolutionaries in Geneva. His action conforms to this description in Nietzsche: as a counter to negation and ennui "the infliction of suffering" is "a magic of the first order, a veritable bait of seduction to life" (XIII, 76).

Conrad conceives of the evil in Razumov's mind, which has been described at length, as that "festering venom and malignity" (XIII, 36) of the weak in Nietzsche. Conrad's view of diabolism is Nietzsche's view of the old morality – a mere superficiality to modern psychologists but religious motives are to be studied as "touchstones" of the soul (XII, 46 -48). In this spirit and to this end, I believe, this twentieth-century novel revives the demonic. Conrad means to illustrate in Razumov's career the diversion of resentment into spiritualized or disguised cruelty, a psychic phenomenon other Russians illustrate, but less fully. After beating Ziemianitch, Razumov is conscious of a "tranquil, unquenchable hate" and longs for power "to hurt and destroy" (p. 37). It is the moment Razumov becomes demonic and undergoes a marked transformation suggesting an encroachment upon the mind by a supra-rational force. He becomes lucid and persuades himself of the rightness of reaction; "some superior power had inspired him with a flow of masterly argument" (p. 41). And when Razumov talks with Haldin again, he has become demonically clever, an alteration that has been glossed on. On one other occasion Razumov is transformed, when he confesses to Nathalie Haldin. The narrator observes that his confession is torn out of him "by a mysterious force." Conrad conceives of these moments, diabolical and divine, one judges, according to this statement in Nietzsche: "Heavy, melancholy men turn lighter, and come temporarily to their surface . . . by hatred and love" (XII, 89). If the narrator does not remark upon Razumov's "melancholy" in the first part of the novel, he gives it great prominence in the second. Some time after beating Ziemianitch, Razumov argues himself into conservatism, as noted above. It is said that "grace" (p. 40) entered his mind in the form of a belief in the autocrat of the future. The passage is puzzling because Razumov's later defection and the wrongness of his actions toward Ziemianitch and Haldin suggest a misapplication of moral terms. The difficulty is resolved if one reads "grace" as a word of Nietzschean analysis and not of Conradian attitude. What the novel illustrates at this point is the bad conscience which in Nietzsche invents hell and the devil and all the supernatural. Here it takes hold of the idea of God and God's autocrat. The novel illustrates, in Nietzsche's phrase, "deified cruelty." The encroachment of Nietzschean literateness upon speech which realistically cannot be so literate creates another difficulty worth mentioning, namely, in that passage in which Razumov remarks of fashionable figures he sees in the streets that they are not capable of "an act of conscience." The statement is true only in a Nietzschean sense that the masters have no conscience, which is an invention of the oppressed.

Many aspects of *Under Western Eyes* remind one of Nietzsche. If Razumov reflects a Nietzschean analysis of the subservient, Peter Ivanovitch and his coterie reflect the psychology of those who subvert by violence but, more importantly, by a transvaluation of values. Madame de S gives to the revolutionary circle in Geneva a bizarre coloring of the spiritual and the mystical. And Peter Ivanovitch does also. His feminism, a cause Nietzsche excoriates, surfaced during his flight across Siberia, as did his animal nature. The "civilised" man and "the wild beast" fled westward under a swarm of flies and mosquitos. Descriptions of the brute in him are Nietzschean. The morbid feelings of guilt because of his chain and his morbid pessimism, further, seem derivatives of the morbid psychology of the slaves. In his writings Peter Ivanovitch treats his escape "mystically" and "symbolically" (pp. 151-155). The second of Nietzsche's two chief objects of contempt, the ascetic and the "desert of the ascetic," is also represented in these two figures. One oblique example is the miserliness of Madame de S, which accounts for the absence of furnishings in the Chateau Borel and for the meager diet there. Secondly, Peter Ivanovitch, when Natalie Haldin seeks him out in his hotel on the night Razumov confesses, has the appearance of a "monk," a "prophet," "a desert-dweller" (p. 410). In his Preface Conrad says that no one in the novel is a monster but that Peter Ivanovitch and Madame de S are "fair game"; they are "apes of a sinister jungle" (p. xii). Peter Ivanovitch is an animal, a mystic, a monster by description and implication, and the high priest of revolution. Surely only Nietzsche could have suggested this final assortment of characteristics. He depicts the priest as a "new animal monstrosity" boar, panther, fox, working against the beasts of prey by guile (XIII, 162-163).

Some small, arresting touches recall Nietzsche – Tekla's "shabby hair and eyebrows" (p. 289), for example, a startling description but one which has its parallel in "the grey and fatalistic mark of the Chandala" in the brows of Nietzsche's inferior men. "The thoughts and actions of all such natures are tainted with a subterranean mouldiness" (XVI, 105). The "Socratic" guise of Councillor Mikulin (p. 111) gives one difficulty until he perceives that "Socratic" in Nietzsche is pejorative and that Conrad uses it of the mild Mikulin in Nietzsche's sense of "life made clear, cold, cautious, conscious, without instincts, opposed to the instincts" (XVI, 16). The novel runs down Geneva and the Genevans insistently, describing the city's artificial vistas with distaste and characterizing its citizens as follows: " 'Elector! Eligible! Enlightened! ' Razumov muttered to himself. 'A brute, all the same' " (p. 254). This outburst repeats Nietzsche's judgment of Geneva and Basle: cities whose temperament is "by frightful mutilations and chronic depressions" turned into its opposite (XIII, 186). Recent criticism has made much of the statue of Rousseau in whose shadow Razumov writes his memoir. The novel does not make plain its significance. But no doubt the Rousseauvean, like the Socratic, is at least partly pejorative. Rousseau, "idealist and *canaille,*" is to Nietzsche an example of a false return to nature (XVI, 60) and a contrast to his own. Probably Conrad intends that Rousseau shall preside as patron over an attempt to penetrate motives with insufficient insight.

II

Appreciation of the Nietzschean quality of *Under Western Eyes* enables the reader to assess the novel's morality much more accurately, usually a matter of distinguishing "Nietzschean" analysis from Conradian attitude. Most prominently the novel offers a moral commentary upon the subject of fate, the subject of autocracy, and the subject of Samaritanism, as it is called.

Razumov's career determines principally the meaning fate has. Events in which he is involved fall into place like Lucretian atoms according to a rigorous causality, one of the novel's most striking effects. A chief example is Razumov's enmeshment in the web of Mikulin's designs. The Councillor foresees accurately that the revolutionaries will approach Razumov and thus supply his bureau with an incomparable agent even as he foresees that Razumov will assent. He calls Razumov an instrument of providence. And Prince K "taken into the secret, was ready enough to adopt that mystical view too" (p. 384). One does not conclude that these transcendental interpretations express Conrad's opinion. The revolutionaries, also, especially Haldin, see providence in some of the same events. The novel emphasizes mysticism as a Russian characteristic, and Nietzschean analysis helps to identify it further as the mark of an oppressed people.

Razumov's reflections upon the betrayal of Haldin offer a

better insight into Conrad's own attitude. The whole novels turns, of course, upon this one episode.

> A man's most open actions have a secret side to them. That is interesting and so unfathomable! For instance, a man goes out of a room for a walk. Nothing more trivial in appearance. And yet it may be momentous. He comes back – he has seen perhaps a drunken brute, taken particular notice of the snow on the ground – and behold he is no longer the same man. The most unlikely things have a secret power over one's thoughts – the grey whiskers of a particular person – the goggle eyes of another. (p. 72)

Razumov acknowledges here to Haldin the mystery of his own nature. He does not succeed in exposing it to himself, but his selection of influences is a Nietzschean selection; and one readily constructs a Nietzschean psychological pattern out of them. Razumov has seen Ziemianitch and beaten him unmercifully. In the snow he has seen an hallucination of Haldin. And he has betrayed him to two members of the ruling class. The whiskers belong to Prince K and the goggle eyes to General T. Hallucination is the Nietzschean sign of the ascetic highly susceptible to the seduction of cruelty. Prince K is a benign figure Razumov associates with autocracy, one who will understand Razumov's "reasonable adherence to absolutism." His influence, in Nietzschean terms, is toward the spiritualization of cruelty. General T possesses the nature of a predatory beast. As a "prey" Razumov feels most strongly at this time that threatening pressure. Consequently, Razumov experiences for Haldin, even as he talks to him, a "murderous" hate (p. 70). The pattern is that of a Nietzschean trauma in which a victim succumbs to the lure of spiritualized cruelty and is transformed. A similar pattern exists in Razumov's agreeing to serve Mikulin. Razumov's transformations set in motion long chains of events. Thus far Conrad's causality is naturalistic.

The novel offers, in the second place, some rather one-sided debate between Natalie Haldin and the teacher of language on the subject of fatality. Contrasting Russia and the West, Natalie Haldin observes to the narrator: "I believe that you hate revolution; you fancy it's not quite honest. You belong to a people which has made a bargain with Fate and wouldn't like to be rude to it. But we have made no bargain" (pp. 166 -167). The bargain the English have made with fate is the exclusion of fate from the body politic. Self-governing, bourgeois societies, evolving according to the "secular logic of human development," have renounced revolution and autocratic rule as instruments of statecraft, thus eliminating artificially those conditions in which fate operates. Political and social life suffers a loss of dimension. Natalie Haldin offers this alternative: "There must be a necessity superior to our conceptions . . . . We Russians shall find some better form of national freedom than an artificial conflict of parties – which is wrong because it is a conflict and contemptible because it is artificial. It is left for us Russians to discover a better way" (pp. 131 -132). Conrad beguiles his reader by putting Nietzschean sentiments in the mouth of a lady. The idiom even is borrowed. Nietzsche writes that the blonde beasts "come like Fate" and have nothing to do with "contracts" (XIII, 103).

Conrad interests himself in causality at least as far back as *Lord Jim,* where he contrasts Marlow's ordered and secure existence to Jim's fateful career. Conrad there recognizes that the individual may choose between kinds of fatality and causality. Even though Jim's freedom can be described as a Nietzschean freedom to obey the will to power, the novel offers no tangible evidence that Conrad conceived Jim's career in that Nietzschean way. One doubts it even more when he considers that *Chance* and *Under Western Eyes,* late works, full of Nietzschean traces, supply no positive illustration of "freedom" but offer instead in their heroes' careers negative illustration of the causality of the ascetic ideal and of spiritualized cruelty. Conrad's late works do not describe an alternative with exactness, but they are not altogether silent on the point. *Chance* looks to a universal peace, the peace of nature, that might be ushered in by ending the universal unrest, created by a false idealism, seething beneath the surface of social order. *Under Western Eyes* looks to a period of "concord and justice" which will end the deepest anguish of the Russian people exemplified by Razumov's betrayal of Haldin. Neither vision suggests that Conrad subscribed to the correlative of Nietzschean fatality, namely, the morality of the beast of prey. A saintly woman, after all, utters the anti-democratic sentiments quoted above.

One last point concerning fatality remains. The harmonizing influence of the presence of Natalie Haldin restores Razumov to inner peace. The narrator observes that a "mysterious force" tears his confession out of him. And at the time Razumov exclaims, "I have the greatest difficulty in saving myself from the superstition of an active Providence" (p. 437). One recognizes that Nietzsche describes the extraordinary force with which love as well as hate possesses the soul of the tamed animal. But still one judges from this episode that Conrad assigns a necessity to that inner harmony he envisions. In *Chance* Anthony and Flora's surrender to instinctive life also has a Blakean inevitability.

Nietzschean analysis clarifies Conrad's political theory, especially with respect to its primary object, autocracy. The novel's most perplexing aspect is the fact that both revolutionists and reactionaries express their beliefs "mystically"; God is everywhere and nowhere. The condition does not exist in Nietzsche's master-slave relationship, where the slaves invent God, but the editor of the English edition of Nietzsche's works applies Nietzschean analysis to contemporary Russia and obtains Conrad's results as follows. The powerful have adopted the religion of the slaves so that "on the Continent everything is muddled in matters of religion." Levy contrasts the naturalness of England to the unnaturalness of Eastern Europe. In England the election slogan is "To Hell with the Dukes and the Lords: vote for Christ! " On the Continent aristocrats "pretend to be on the most intimate terms with Him" (XVIII, xv).

Razumov argues himself into a belief in the autocrat of the future. His argument is suspect because it is rationalization and a part of the Nietzschean psychology of the deification of cruelty. But Natalie Haldin, also, looks to a benevolent autocrat. Her beliefs, formed under the same tyrannical pressures as Razumov's, are also suspect but are not so easily dismissed. The novel's advocacy of autocracy will not die under "Nietzschean" analysis. The sane and attractive side of the two characters contributes to the formation of their beliefs; and, of course, Nietzsche himself writes of the "blessings" of autocracy and of its being indispensable to the "herd mentality." Conrad seems drawn to Nietzsche's prophecy of the coming of an "Antichrist and Antinihilist" (XIII, 117). Conrad, sharing Nietzsche's distaste for parliamentary forms, allows Natalie Haldin to express it. Furthermore, she echoes Nietzsche's wish (XIII, 23) that "one could invent something new in politics." Natalie Haldin's splendid assertions of the moral superiority of Russia over Western Europe closely parallel these passages in Nietzsche: "I observe more inclination towards greatness in the feelings of the Russian Nihilists than in those of the English Utilitarians" (XIII, 223) and Nietzsche calls Russia "the *only* great nation to-day that has some lasting power and grit in her, that can bide her time, that can still promise something. – Russia the opposite of all wretched European petty-statism and neurasthenia" (XVI, 96-97).

Conrad's favorable attitude toward autocracy in the novel is not recent. *The Secret Agent* expresses distaste for modern plutocracy. *Nostromo* pictures modern democracy dominated by material interests destroying the fidelity of a simple soul. Lord Jim is a benevolent autocrat and so, of course, is a ship's captain, such as Conrad once was. In *Under Western Eyes* Conrad conceives of a social order different from that he knew in England, but his views on the subject are not more definite or systematic than Nietzsche's. Conrad is not primarily a political theorist.[11]

The example Tekla and Natalie Haldin set by devoting their lives to the most wretched among the Russian people may seem an admonition to extreme altruism, but the admonition fades when the reader scrutinizes it closely through a Nietzschean lens. Where the nihilistic prevails, Nietzsche writes, only two things flourish: "the metapolitics of St. Petersburg and the 'pity' of Tolstoi" (XIII, 203). Nietzsche rails against *"la réligion de la souffrance* humaine" (XII, 31); altruism is decadence or a "phase of cruelty" (XIII, 105); Nietzsche makes much of the feeling of superiority men derive from charitable acts (XIII, 175). In spite of the attractiveness of Natalie Haldin, the women's action conforms so completely to a predicted course that the reader must conclude it results from analysis rather than attitude. What value it has as admonition is implicit in the following discussion.

The novel does erect a moral standard through assumptions evident in certain of its pronouncements, as in this passage, which enumerates what Russia is not: "the noblest aspirations of humanity, the desire of freedom, an ardent patriotism, the love of justice, the sense of pity, and even the fidelity of simple minds are prosti-

11. "Organicism," as Avrom Fleishman uses it, is a very broad term; but I do not believe that it can encompass the Nietzschean attitudes I have described. Our interpretations are essentially different. See *Conrad's Politics*, pp. 76-77, 217-42.

tuted to the lusts of hate and fear" (p. 6). The passage intimates that generous impulses, developing normally and not abnormally under the pressures of tyrannical lawlessness, are the natural endowment of men. As the narrator observes, they are the birthright of English youth, who cannot imagine an existence in which it is possible to be imprisoned on suspicion and to be whipped. Razumov apologizes, further, for his murderous acts by citing his orphaned condition. The educated reader in 1910 would have understood Conrad's meaning. Utilitarian moralists of the nineteenth century and early twentieth century habitually assert that the family is the origin of utilitarian and altruistic values. The modified Spencerian and Utilitarian morality in Conrad's fiction dating at least from *Heart of Darkness*[12] survives in this novel; and the fact testifies to Conrad's respect for the moral and psychological stability of the English mind. But it is not the complete morality of *Under Western Eyes.* The superiority of East to West is maintained, finally, with respect to personal temperament. Before he betrays Haldin, Razumov considers rushing to him and confessing his treacherous thoughts. He would thereby initiate a "fellowship of souls – such as the world had never seen" (p. 47). In this case brotherhood would exist between men who had known and rejected the lure to cruelty. This altruistic action has more dimension than the Spencerian. Conrad seems to follow the editor of the English edition in believing, in Nietzschean terms, that to be good, one must first become evil (XVIII, xxiiiff.). Certainly, the statement summarizes the whole of Razumov's career.

A brief comment should be offered, in conclusion, on Conrad's purpose. One decides that, above all, he intends in Nietzsche's phrase to compare "kinds" or "types" of morality (XII, 103-104). Nietzsche observes that morality hitherto has been something "given" – and now there are no moral absolutes. Conrad is not as emancipated, but his purpose is comparison and his choice of subject seems dictated by Nietzsche's criticism of

12. See my articles "Spencerian Assumptions in Conrad's Early Fiction," *Conradiana,* I (1969), 29-40; "Contemporary Psychology in 'The Nigger of the Narcissus'," *Studies in Short Fiction,* VII (1970), 219-31; "Conrad's View of Primitive Peoples in *Lord Jim* and *Heart of Darkness,*" *Modern Fiction Studies,* XVI (1970), 163-83; "Marlow's Psychological Vocabulary in *Lord Jim,*" *Texas Studies in Literature and Language,* XII (1970), 457-70.

English and French "sociology" for knowing only "the decadent form of society" (XVI, 93). The "moral discovery" of *Under Western Eyes* is discovery of those moral phenomena indigenous to a master-slave society, surviving in modern Russia.

PART TWO: CHAPTER TWO

The Morality of *Chance*

The Morality of *Chance*

Conrad wrote to Galsworthy that he composed *Under Western Eyes* as a respite: "I had to get away from *Chance*, with which I was making no serious progress."[1] The remark leaves an impression that coping with the subject of *Under Western Eyes* aided Conrad eventually in resolving whatever problems the subject of *Chance* posed. I have argued that *Under Western Eyes* is a novel of Nietzschean analysis.[2] The relationship Conrad describes between the two novels makes inevitable the surmise that *Chance* is also Nietzschean. The Nietzschean insights of *Under Western Eyes* are primarily political, but important to that novel also is Nietzsche's polemic against the ascetic ideal. I believe that in *Under Western Eyes*, almost incidentally, Conrad found that Nietzschean direction with respect to love and followed it with increasing absorption through his last three major works of fiction.

In *Chance*,[3] from the beginning of their marriage, Captain Anthony absolves Flora de Barral from having a sexual relationship with him; and that is a great renunciation, for he has a passionate nature. Fyne, sent to London by Mrs. Fyne to break off relations with her brother because of the unsuitable marriage, brings about the abnormal condition. He reveals to Anthony the motive Flora inaccurately expressed in a letter to Mrs. Fyne: she will marry only to gain Anthony's protection. True to a kind of moral subtlety of which the Fynes – and Captain Anthony – are capable, Little Fyne accuses Anthony of marrying for sexual pleasure a woman who does not love him. As Marlow points out, Anthony is the son of an esteemed poet who made "touching, delicate, fascinating the most hopeless conventions of the so-called refined existence" (p. 224). In the spirit of Victorian idealism,

1. G. Jean-Aubry,. *Joseph Conrad: Life and Letters* (London: William Heinemann, 1927), II, 65.

2. Chapter One above appeared as an article in *Criticism*, 14, No. 1 (Winter, 1972), 32 -48.

3. References to *Chance* in my text are to *The Works of Joseph Conrad* (London: William Heinemann, 1921), XIII.

Anthony acts altruistically; and Flora, the passive victim of a long series of hostile acts, passively accepts his decision because she believes that no one can care for her. When the Pyrrhic marriage becomes almost insufferable, old de Barral's attempt to poison Anthony causes Anthony and Flora to disclose their real feelings and brings about a normal relationship.

The unusual relationship between Flora de Barral and Captain Anthony is of a chivalric kind such as Nietzsche condemns. Anthony's role – it is said of him that he was "intoxicated with the pity and tenderness of his part" (p. 302) – Nietzsche describes under the classification of "ascetic idealism in the service of projected emotional excess":

> This secret self-tyranny, this cruelty of the artist, this delight in giving a form to one's self as a piece of difficult, refractory, and suffering material, in burning in a will, a critique, a contradiction, a contempt, a negation; this sinister and ghastly labour of love on the part of a soul, whose will is cloven in two within itself, which makes itself suffer from delight in the infliction of suffering; this wholly *active* bad conscience has finally (as one already anticipates) – true fountainhead as it is of idealism and imagination – produced an abundance of novel and amazing beauty and affirmation, and perhaps has really been the first to give birth to beauty at all.[4]

Nietzsche's descriptions fit also the whole relationship and summarize its course:

> There are proceedings of such a delicate nature that it is well to overwhelm them with coarseness and make them unrecognisable; there are actions of love and of extravagant magnanimity after which nothing can be wiser than to take a stick and thrash the witness soundly: one thereby obscures his recollection. (XII, 55)

The novel's intolerable relationship is "overwhelmed" at last by violence and by candor, probably, that is to say, weighing the difference between modern and Victorian attitudes, by "coarseness." Correspondence between Conrad's images and Nietzsche's abstractions is precise.

The novel has its own abstractions, and these correspond to

4. Oscar Levy, ed., *The Complete Works of Friedrich Nietzsche* (London: T.N. Foulis, 1910), XIII, 105.

Nietzsche's. The following passage from *Chance* is a commentary on the thwarting of physical love by idealistic impediments:

> And the punishment of it is an invasion of complexity, a tormenting, forcibly tortuous involution of feelings, the deepest form of suffering from which indeed something significant may come at last, which may be criminal or heroic, may be madness or wisdom – or even a straight if despairing decision.(p.496)

Like Nietzsche, Conrad finds idealism the cause of enormous psychological complexity and of suffering and delight; even though suspect, it is the origin of actions which can be called wise, affirmative, beautiful. Like Nietzsche, Conrad offers not a conventional but a searching and complex insight, expressed ironically with seeming contradictions. And the tone taken, between anguish and exaltation, is similar.

Beyond these correspondences, what convinces the reader that Conrad's conception is Nietzschean is the fact that Marlow uses Nietzschean words and phrases to set it forth. Words such as *ideal* and *idealism*, expletives such as *mad* [Anthony's renunciation is a "mad" action; his is the "madness of generosity" (p. 388)] are common to both Conrad and Nietzsche and, when applied to the same object, establish a broad similarity. But the conclusion that Conrad's analysis is Nietzschean results more importantly from the presence in *Chance* of Nietzschean words with more specialized meanings and stronger connotations. Two of these occur in the passages from Nietzsche quoted above, i.e., *magnanimity* and *delicacy. Magnanimity* ordinarily names a larger and more intellectual altruism than the spontaneous; in the tradition of Edmund Spenser it is the sum of all the virtues, and is according to Nature; in Nietzsche's very specialized usage *magnanimity* names a complexly motivated action thwarting natural instinct. Conrad reaches out frequently beyond "generosity" for the more meaningful word; in the following examples the word conveys the distinctly Nietzschean irony: Flora must face "this new perfidy of life, taking upon itself the form of magnanimity" (p. 389); there is a perfection in his "magnanimity" [Captain Anthony's] "which makes it almost impossible to bear" (p. 451); Anthony cannot face de Barral's attempt to poison him "for the sake of his mere magnanimity. The normal alone can overcome the abnormal" (p. 499).

Delicate is a Nietzschean word, describing in a passage above idealistic inventions. It has a further resonance in that Nietzsche used it to describe the priest in his role with respect to the transvaluation of values. Nietzsche calls him "the first form of the more delicate animal that scorns more easily than it hates" (XIII, 162). The inventions of the delicate animal will mortify natural instinct and gratify subtly the craving for superiority, which is a modified form of the will to power. The word is used with respect to Captain Anthony in an unadulterated Nietzschean sense. Marlow says Anthony's heroic resolve "wears the aspect of sublime delicacy " (p. 379), that is to say with Marlow and Nietzsche, of sublime idiocy. Anthony's sister also has "delicacy" (p. 189). In her youth she suffered with Anthony the tyranny of a father whose temperament contained both poetry and ferocity. By virtue of the fact that she was one of the repressed, she might be expected to be susceptible in a Nietzschean way to the lure to idealism. Conrad devised for her, however, another development. She married the very "masculine" civil servant Little Fyne. Marriage to Fyne is salutary and restorative. His health, Mrs. Fyne's health, and the health of their red-cheeked daughters is transparent. In contrast to his early judgment of Anthony, Marlow's judgment of Mrs. Fyne is that she possesses "unflinching steadiness" in a "pure compassion" (p. 161); she has "None of your dishevelled sentiment" (p. 163). In the English and Spencerian tradition of altruism, her delicacy is the refinement of the generous impulses of a superior individual. As such, they fall short of the complex "delicacy" of Nietzsche's idealist or of his perfected moralist. In the sense of this insufficiency, no doubt, Marlow's use of *delicacy* is playful. However, Mrs. Fyne's intimate perception of Anthony's temperament through shared experience separates her from Fyne and makes her the leader in their negotiations. Possessed of that additional meaning, *delicacy* is not only playful but ambiguous and conveys by indirection the Nietzschean complexity.

A third word is *transcendental*. It is used only once, I believe, in *Chance.* But like *magnanimity* it is an unusual and arresting word. And like *magnanimity* it is used in *Chance* not in a familiar but in a Nietzschean sense. Surely Conrad's "transcendental good intentions, which, though ethically valuable . . . cause often more unhappiness than the plots of the most evil tendency"

(p. 436) are Nietzsche's "transcendental aspirations, contrary to sense" (XIII, 116).

A final word is *egoism.* It is not in the same way the others are a Nietzschean word. Nietzsche contrasted his morality, the analysis of motives and the recommendation of the will to power, specifically to English Utilitarianism. Hence in philosophical discussions in England at the turn of the century, as noted in the preceding chapter, the important opposition was between the traditional "altruism" and a challenging "egoism." In highly literate moral discourse in English at this time, *egoism* signifies much or all of Nietzschean philosophy. The meaning of Marlow's insistence on the "semi-conscious egoism of all safe, established existences" (p. 177) and "the instinct of self-preservation and the egoism of every living creature" (p. 206) should be decided in the light of that fact. This vein of observation is prominent in the novel, resembling a first principle like the Nietzschean principle of the prevalence of the will to power.

Chance casts more suspicion upon altruistic motives than any other novel before *Under Western Eyes.* Toward the end of *Chance* Marlow observes that "A residue of egoism remains in every affection – even paternal" (p. 429). In that skeptical climate the most unshakable truism is shaken, or stated in a qualified way, for example, the following: "Generally speaking, an unselfish action is a moral action" (pp. 221 -22). As has been suggested, acts of charity to Nietzsche are disguised manifestations of the will to power; they bestow a feeling of "superiority" (XIII, 175). That unusual view must stand behind the novel's unusual characterization of pity as being "akin to contempt" (p. 383). Nor is love exempt. A Nietzschean analytical complexity and pervading skepticism, wholly un-English, informs Marlow's remark "If Anthony's love had been as egoistic as love generally is, it would have been greater than the egoism of his vanity – or of his generosity, if you like" (p. 383).

II

A second intimation of Nietzschean influence is Conrad's use in *Chance* of animal metaphor with consistency and good sense. Conrad used animal metaphor in his earlier fiction, but that it often lacked the qualities just named can be seen in the following, fairly typical examples. In *Lord Jim* Marlow likens the Captain of the *Patna* to a "trained baby elephant walking on hind legs" (IV,44); and only a few pages later Marlow asks, "What decent little port on the shores of the Baltic was defiled by being the nest of that precious bird" (p. 50). *The Secret Agent* contains images which are the work of fancy and depend upon superficial likenesses – for example: "Speaking tubes resembling snakes were tied by the heads to the back of the Assistant Commisioner's wooden armchair, and their gaping mouths seemed ready to bite his elbows" (VIII,109). Toward the end of the novel, it is said, the "gas jet purred like a contented cat" (p. 262). These images are fortuitous; in contrast, the images of *Chance* cohere and convey impressions of human nature which are evolutionary and, I believe, Nietzschean.

They convey, in the first place, a general impression of animal nature in human nature. The animal looks out of the human form. The most obvious example occurs in a passage in which Marlow catches Fyne in an "extraordinary simiesque grimace." Marlow reflects that a "grave, much-lined, shaven countenance, when distorted in an unusual way, is extremely apelike" (p. 281). Further, Marlow says that Mrs. de Barral "died as some faithful and delicate animals die – from neglect" (p. 82). The simile implies a temperamental state surviving from an animal condition. Or man shades into the animal more explicitly when Marlow calls him a "reckless animal," one "so made that in him curiosity, the paltriest curiosity, will overcome all terrors" (p. 55).

The most striking impressions in Nietzsche of the animal-man are of a creature mad, melancholy, and cruel. These traits have not been obvious in the images cited, but they reside in other metaphors in *Chance.* One of the earliest animal metaphors in the novel uses Nietzsche's most characteristic phrase. Dockside ruffians Powell encounters on his way to the *Ferndale* are said to be "wary in their movements and perfectly silent of foot, like beasts of prey slinking about a campfire" (p. 29). In a meeting of Marlow and Flora at the time the Fynes are working to prevent Flora's marriage, Marlow says he cannot tell whether he sees in her "a smile or a ferocious baring of little even teeth" (p. 234). When Anthony tries to persuade Flora to go with him in the *Ferndale*, it is said with some ambiguity that he "growled at her in a savage passion" (p. 262).

By far the most interesting metaphor of this kind describes the relative of the de Barrals as he takes Flora from the Fynes: he "fastened his fangs, figuratively speaking, into the poor girl and prepared to drag her away for a prey to his cubs of both sexes" (p. 195). The cousin's charity is a form of cruelty, for they torment Flora in sly and sanctimonious ways; Marlow refers to their "moral savagery" (p. 189). The beast-of-prey metaphor in this example reveals animal ferocity under civilized usage and idealistic behavior, and that is a Nietzschean kind of insight. The effect of the example is broad. By identifying the unconscious motive in the actions of these ambivalent characters, it implies the same motive in the actions of others: of Carleon Anthony, a poet who exhibits "traces of the primitive cavedweller's temperament" (p. 44); of Captain Anthony, who late in the novel "acts like a beast of prey" (p. 483); of the murderous de Barral, and the "savage" governess.

The governess obviously desires power; her temperament is unsuited to her situation in life. The novel phrases that disharmony with great effect; she has endured "bitter years of restraint"; she has cultivated an "iron, admirably mannered restraint"; exercise of that restraint is "like living half strangled" (p. 139). In her case Conrad appears to represent the plight of uncommon individuals overcome by a Nietzschean slave morality. The governess eases her tension by hypocrisy, "a way of getting even with the common morality" (p. 138); but she gratifies herself most completely by her devastating attack on Flora. The episode seems written in the spirit of Nietzsche's descriptions of Europeans who as adventurers achieve freedom from social control; "they revert to the innocence of the beast-of-prey conscience, like jubilant monsters" (XIII, 40). The governess' assault defines the nature of evil in *Chance.* Marlow describes that action in generic terms as that "evil which lives in the secret thoughts and therefore in the open acts of mankind, whenever it happens that evil thought meets evil courage" (p. 115). It is by implication the cause of human disquiet and the planet's unrest.

Another set of animal metaphors renders not the cruelty of the beast of prey but the terror of its victim.Flora's first reaction to the last visit of the governess' accomplice is phrased as follows: their look "reached, stirred, set free that faculty of unreasoning explosive terror locked up at the bottom of all human hearts and of the hearts of animals as well. With suddenly enlarged pupils and a movement as instinctive almost as the bounding of a startled fawn, she jumped up" (pp. 134-35). Flora's response later to her new status, in context, conveys a similar sense of animal terror: "she started into the full consciousness of her position, like a person waking up in contact with something venemous – a snake, for instance – experiencing a mad impulse to fling the thing away and run off screaming to hide somewhere" (p. 210). Conrad applies to her the image of a trapped animal [Flora is "like a creature struggling under a net" (p. 162)] and to two other "victims." De Barral reacts to the discovery of his attempt to poison Captain Anthony as follows: "Like some wild creature scared by the first touch of a net falling on its back, old de Barral began to struggle. . . with staring eyes and gasping mouth" (p. 423). Powell reports of Anthony, defeated by the situation on the *Ferndale*: "I saw him suddenly fling his head to the right and to the left fiercely, like a wild animal at bay, not knowing which way to break out" (p. 491).

Many of these passages, like some cited at the beginning of this section, go beyond metaphor. Their effect is that of analogy, and they are accompanied by explicit analysis. In them Conrad expends his imaginative energies lavishly to create an extraordinary psychology of human terror from the observation of animal nature. The imaginative concept, one judges, is original; but some of its originality consists in following two suggestions from Nietzsche. In the first place, women in Nietzsche are especially disadvantaged. They are "sufferers of the lower orders"; a woman is "a compound of labour-slave and prisoner" (XIII, 175). In Nietzsche repression dams up animal and savage traits, which break out irresistibly. Women, being most repressed, can be expected to be most savage; Nietzsche says as much. That Nietzschean argument surely accounts for the preeminence in evil of the savage governess. Flora's lot is most like the lot of women Nietzsche describes; she is literally "labour-slave and prisoner"; but she is not a jubilant monster. Apart from his own conviction, Conrad would have recognized the hyperbole in some of Nietzsche's analyses of women. In her case what stimulated Conrad's imagination most likely is the general sense one gains in the reading of Nietzsche of the oppressed falling back upon animal reserves – animal cunning, for example – and, beyond that, of this specific characterization

Nietzsche gives to the slaves: they suffer "an enervation of the mind and relaxation of the limbs" (XIII, 37). The two impressions combine in this stunning passage, describing further Flora's reaction to the governess' assault, in which an extreme threat upon a timid mind awakens animal alertness, intense apprehension, preternatural quiet: "Flora had to know all the possible phases of that sort of anguish, beginning with instinctive panic, through the bewildered stage, the frozen stage, and the stage of blanched apprehension, down to the instinctive prudence of extreme terror – the stillness of the mouse" (p. 140).

III

Chance is the most playful of Conrad's novels. Marlow, who has become a bantering seer, offers "practical sagacity" to a leader of the feminist movement, whom a younger lady has distressed by turning the older lady's theories against her. The younger lady, who is not acceptable as a sister-in-law because her father is in jail, proposes to secure her position in a world dominated by men by marrying the older lady's brother. Marlow in this novel is a complete moralist. He is not ambivalent or hesitating or being instructed by experience as in *Lord Jim* or *Heart of Darkness.* He completes analyses and renders judgments and from the vantage point of his comprehensive knowledge of human nature is amused especially by the unconscious motives of men.[5] What is most playful in this, I believe, is that the seer's practical sagacity and assurance derive from Nietzsche, who was antagonistic to the feminist movement and whose theories in 1912 must have seemed brutal to most ladies. *Chance* is more than a little malicious; but Conrad must be forgiven because in the end he pays feminine nature a supreme compliment. He makes out that the fatal inclination of the race toward idealism is a masculine rather than a feminine penchant. Men are responsible, in the words of the novel, for the "complicated bad dream of existence." Little Fyne drives Anthony into idealistic folly, and that is why the

5. J.W. Johnson expresses a similar opinion in "Marlow and *Chance:* A Reappraisal," *Texas Studies in Literature and Language,* 10 (1968), 91 -105. See also Robert N. Hudspeth, "Conrad's Use of Time in *Chance,*" *Nineteenth-Century Fiction,* 21 (1966), 283 -89.

novel insists on Fyne's "masculinity." Mrs. Fyne is also "masculine," Marlow says. She has lost her feminine insight by becoming a ruthless theorist and feminist, that is, idealist. The hope of the world lies in women's usual acuteness, their lack of ideation with respect to the objects of everyday. The novel commends even the despicable governess on that count; she is "uncommon enough to live without illusions" (p. 121). Thus, Marlow's first recommendation as a moral philosopher is a recommendation to a new realism.

His second recommendation is to a new humanism to which that realism is conducive. The two things are related in the following passage, a quintessential characterization of Marlow: there was "in his eyes that slightly mocking expression with which he habitually covers up his sympathetic impulses of mirth and pity before the unreasonable complications the idealism of mankind puts into the simple but poignant problem of conduct on this earth" (p. 376). Compassion here originates clearly enough in a Nietzschean kind of insight; the statement is a mild polemic against the ascetic ideal. One other passage says the same thing. Marlow observes that nature provides mercifully that people following their illusions do not know what is happening to them, that other people do understand and "from that same provision of understanding there springs in us compassion, charity, indignation, the sense of solidarity; and in minds of any largeness an inclination to that indulgence which is next door to affection" (p. 136). The passage raises the question, how large must "minds of any largeness" be? The answer the novel supplies is, first of all, much larger than the minds of Flora's middle-class relatives, whose charity is barely unconscious cruelty and represents the common morality. In the second place, the answer is larger than the minds of the two Fynes, those highly evolved creatures, whose compassion has merit. In an interesting analysis Marlow says they lack "imaginative sympathy" (p. 165), which is true only in the sense they have not experienced isolation as Flora and Marlow have; they have not reflected upon human nature or come to complete self-knowledge in alienation or in the solitude of the sea. They lack that pity which is not akin to contempt felt by "silent solitary men" (p. 383) such as Anthony . Conrad here seems to distinguish between a Spencerian kind of empathy and a compassion associated with a higher degree of self-knowledge, conceived out of his own experience with significant inspiration from Nietzsche. Conrad's distinction is like that Nietzsche draws between English altruism and the position of his own moralist beyond good and evil; it is reminiscent of Nietzsche's assertion that to be good one must first be evil, that is, the true moralist arrives at wisdom by experiencing and freeing himself from a slave morality.

The felicitous life which Conrad's moralist recommends has three or four aspects significant enough to be singled out for further comment. Some of these diverge from Nietzsche. In the collapse of the ascetic ideal Conrad does not espouse the morality of the blonde beasts. An imperative to instinctive action remains common to both, but a difference exists in quality and tone. In *Chance* such instinctive action is manly and beneficent; it does not destroy, it enhances "solidarity." It is an action of innocence: "And if two beings thrown together, mutually attracted, resist the necessity, fail in understanding, and voluntarily stop short of the – the embrace, in the noblest meaning of the word, then they are committing a sin against life, the call of which is simple perhaps sacred" (p. 496). Conrad's sense of necessity in the call of life is strong and is the reason for Marlow's perpetually pondering whether the astonishing patterns in individual lives come about by chance, design, or providence. Marlow's use of "chance" throughout suggests not chance happenings but happenings by cosmic design to bring about an individual's involvement in life. Those who obey the call of life achieve the "serenity of the unconscious nature," and Marlow has this optimistic vision: "it seemed to me that it must reign everywhere, over all the globe of water and land and in the hearts of all the dwellers on this earth" (p. 514). In Nietzsche, on the other hand, the imperative to instinctive life is complex, self-centered, and violent. Nietzsche would free his Overman from the causality of a bourgeois society, but there is no necessity; nor is a world where the blonde beasts descend "like Fate" serene.

On one final point Conrad and Nietzsche agree entirely. In both, as has been seen, a painful complexity invades those who fail to act instinctively. In characterizing the governess, Marlow adds the penalty of despair. He speaks of the "pessimism of the passionate who at bottom feel themselves to be the outcasts of a morally restrained universe" (p. 123). He has enough sympathy for her plight – as Nietzsche has for the blonde beast entrapped by slaves – to observe that her practice of hypocrisy is "a way of getting even with the common morality from which some of us appear to suffer so much" (p. 138). This vein of analysis amounts to a Nietzschean critique of the origin of pessimism in the ascetic ideal. Nietzsche directed his own against Wagner and Schopenhauer.

Realism and humanism were aspects of Nietzsche's philosophy that about 1912 attracted the attention of English intellectuals, their interest being stimulated by the appearance of the English translation of his works. In the *English Review* reaction to Nietzsche along these lines was most favorable. This similarity can be offered, along with vocabulary and animal metaphor, as an argument for Nietzschean influence on the conception of *Chance*. In the *Review*, "The Theatre of the Future," an article by Viacheslaf Ivanof, translated by Stephen Graham, picks out a Nietzschean trend to realism in modern literature and lauds it as the direction of the future: "Last of all in the realm of philosophy, the reaction against customs and methods of thought tells of the conquest of idealism itself, and the tendency towards primitive realism. Not Nietzsche alone felt himself more akin to Heraclitus than to Plato."[6] Secondly, Horace B. Samuel, who translated *The Genealogy of Morals* in the collected edition, in another article in *The English Review* entitled "The Genealogy of Morals" describes enthusiastically the enhanced possibilities of life opened up by Nietzsche's analyses: "Nietzsche is capable of constituting a valuable modern bible for the twentieth-century man who proposes to live vastly and to play for grand stakes."[7] Nietzsche's existential ideal of conduct, as Samuel summarizes it, has the simplicity of Marlow's: Nietzsche "demurs vigorously to almost any solution of the riddle of the world. According to his reasoning the need for any solution at all, whether transcendental, after the pattern of Kant and the Idealists, or quasi-transcendental, after the pattern of the pseudo-metaphysics of the scientists, argues an inability to take life on its own merits and on its own valuation" (p. 142).

By characterizing *Chance* as a story told from the wrong

6. 10 (1912), 638.
7. 9 (1911), 143.

point of view and its narrative method as doing a thing in such a way as to "make it undergo most doing," Henry James from the beginning cast doubt upon the novel's accomplishment. "Obscuration" and, also, a "noble sociability of vision" were both unfortunate results of using Marlow and the other narrators.[8] Ian Watt has defended *Chance,* arguing that Marlow is a "recording consciousness" in a Jamesian sense.[9] What has been said in this study allows one to speculate that James judged the novel's heavy narrative apparatus with reference to the development of a story line; that James was not temperamentally equipped to appreciate subtleties in "primitive emotions," as Conrad once labeled his purview, contrasting it to James'[10]; that the novel's contemporaneity and erudition escaped him; and that James, therefore, underestimated the novel's morality.

James refers critically to Marlow's "prolonged hovering flight of the subjective over the outstretched ground of the case exposed" (p. 204). James did not discern that Marlow hovers there to study the perverse complexities of the ascetic ideal. Through other narrators, Marlow has as much of an overview as the philosophical narrator of *Tom Jones,* and Marlow has the immediate insights of an involved character. The overview is advantageous in a didactic and philosophical novel; it conveys a comprehensive set of psychological insights and also a developed tenderness. The "I" narrator serves the latter purpose especially. He is more detached from events than Marlow. He characterizes Marlow as a contemplative and thus helps to throw into relief Marlow's humanistic wisdom and recommend it to the reader. James acknowledges Marlow's stance as a philosopher but does not credit him with a philosophy.

8. See Henry James, *The Art of Fiction and Other Essays* (New York: Oxford University Press, 1948), pp. 202-06.

9. See "Conrad, James, and *Chance*" in *Imagined Worlds,* ed. Maynard Mack and Ian Gregor (London: Methuen, 1968), pp. 301-22.

10. *L & L,* I, 270-71.

PART TWO: CHAPTER THREE

Conrad as Moralist in *Victory*

Conrad as Moralist in *Victory*

The weight of critical opinion has gone against Conrad's late novel *Victory,* adverse judgments having been handed down for reasons of both content and form.[1] Although this study aims at a reassessment on both counts, it interests itself especially in the former. It seeks to describe with exactness Conrad's character as a moral psychologist. It asserts against Conrad's detractors that his character in this novel reached its highest development and that *Victory* is a work of wisdom.

Two psychological influences are important in *Victory:* the English associational psychology of the period and Nietzsche's philosophy. The most striking effect of the first is the symbolic characterization of the cat-like Ricardo and the ape-like Pedro. Some critics have acknowledged Conrad's symbolism but without taking it seriously enough or reading it accurately; some others have complained that these pieces of characterization are unconvincing; and, of course, they are, as all symbolic characterizations are in some degree unconvincing. Pedro and Ricardo, even though they are vividly realized images, are not whole men. They are parts of men, aspects of human nature which a turn-of-the-century reader must have recognized at once as the *tiger* and the *ape.* Henry Maudsley, the leading abnormal psychologist, gives their significance sufficiently in this passage: "Destroy the social structure of a nation, as in the French Revolution, and then behold what monsters of apish deformity and tigrish ferocity the individuals are capable of becoming."[2] In some other psychologists the ape is lascivious. The passage from Maudsley conveys the most characteristic insight of contemporary psychology, namely, its awareness of the animal in man, the brutish and the predatory. J.A. Hobson, writing on "The Task of Realism" in 1909 in *The*

1. Albert J. Guerard's opinions have had great authority with other critics. See *Conrad the Novelist* (Cambridge: Harvard University Press, 1958), pp. 272-78.

2. Henry Maudsley, *The Pathology of Mind* (London: Macmillan, 1895), p. 32.

English Review, is much aware of that achievement of nineteenth-century thought; he describes a "new conception of continuity secured by the acceptance of the Darwinian hypothesis of the origin of species, and especially the bridging of the chasm which hitherto had separated man from Nature, with the finer application of causality to the innermost life of man" (III, 545). The passage describes the nature and purpose of *Victory.* It is a "finer application of causality to the innermost life of man," a part of which is figured symbolically by Pedro and Ricardo. Causality is traced, moreover, in a condition of life like that Maudsley envisions, achieved in this instance by the isolation of characters rather than by the historical collapse of social institutions. Maudsley is convinced that "human nature needs the continual support of good social conditions to hinder it from running back to barbarism" (p. 81).

Also important to characterization in *Victory,* although understated, is that phenomenon associational-Utilitarian psychologists call "mobility of mood," a rapid alternating of emotions, most commonly of tenderness and hate. To a point such mental activity obeys a law of our nature, which Cesare Lombroso, as noted in Chapter I, phrased as follows: "Cela est dû à ce que notre moral comme nos nerfs est sujet à la loi des contrastes: une fois la bonté epuisée, on incline vers la méchanceté; une fois la cruauté satisfaite, on incline vers la bonté, de même que la rétine fatiguée du rouge voit du vert, et *vice versa.*"[3] But extreme mobility of mood is a malady which psychologists attribute to an impaired nervous structure in a Spencerian sense. It is in part a too-rapid response to sense stimuli, a "quick, unbridled and dissolute reflex action of cerebral tracts," to borrow a phrase from Maudsley (p. 334). The complex nervous structure of a civilized man and his developed powers of imagination inhibit response and lengthen transitions from emotion to emotion. Impulsiveness and instability are characterics of savages and, accordingly, mobility of mood is thought of as being atavistic and hence degenerate, for contemporary psychologists define degeneracy in terms of regressive characteristics. Théodule Ribot, a widely read French psychologist, takes essentially that view of it when he classifies mobility

3. Cesare Lombroso, *Les Anarchistes* (Paris; E. Flammarion, 1896), p. 149.

of mood as "infantilism" and notes that it is familiar in genius and "grownup children."[4] Such phrases very accurately describe the case of the retarded Stevie in *The Secret Agent.*

Ribot offers as an example of mobility of mood the troubled affair of George Sand and Alfred de Musset, who treated his mistress alternately as a divinity and an object of contempt (p. 421). It was as if a tender soul and a murderous soul warred for the privilege of animating Musset's body. In much of his fiction Conrad used principles of pre-Freudian psychology as fully as later writers have employed the Freudian. Ribot's example illustrates how firm a compact Conrad has in *Victory* with the formal psychology of his time. For Conrad conceives of the relationship between Heyst and Lena in the same way. He phrases Heyst's mobility of mood early in the relationship on Samburan as follows:[5] "In his soul and in his body he experienced a nervous reaction from tenderness. All at once, without transition, he detested her. But only for a moment" (p. 264). The passage conveys in addition to the analysis above the faintest suggestion of Eduard von Hartmann's principle of the "pain of compassion," a sensibility which develops in the most highly evolved members of the race. Heyst calls himself a "man of the last hour" (p. 440).

One other character in *Victory* displays Heyst's symptoms, namely, Ricardo. After assaulting Lena unsuccessfully in Heyst's bungalow, he is affected by Lena's silence, which "tacitly condoned the offence" (p. 361), so that he "already felt almost tender toward the girl." Rapacity here gives way to tenderness, but the novel does not underscore the pathological nature of the phenomenon. It does so later, however, when Conrad describes Ricardo's state just before the desperados make their attempt on Heyst on the night of the storm. Of his need to "rip someone up" and of his affection for Lena, Conrad observes, "A thirst of blood was allied in him with a thirst for tenderness – yes tenderness" (pp. 411-412). There is no mistaking the abnormality or Conrad's reliance upon a psychological principle not accessible to the common reader he addresses. Why should Conrad give to such unlike characters as Heyst and Ricardo the same pathology? The answer,

4. *The Psychology of the Emotions* (London: W. Scott, 1897), p. 422.

5. References to Conrad's fiction are to *The Works of Joseph Conrad* (London: William Heinemann, 1921); *Victory* is volume XV.

I believe, is that they are meant to be one and the same person; or, more accurately, Ricardo represents primitive predatory instincts not only in man but particularly in Heyst. The symbolism of *Victory* is so deliberate that it may accurately be called allegory.

More things in the novel than the fact that Heyst and Ricardo both suffer from mobility of mood lead to the conclusion of their identity. The novel contains numerous signs and signals of the desperados' unreality and, hence, of their symbolic value. Conrad, for example, likens their arrival to the arrival of strangers in Polynesian legends and so places them for the reader in the world of myth. More particularly, Wang's reaction to the assault on Lena, in which he seems to assume or is assumed to assume that the attacker is Heyst, even though later on he knows it was Ricardo, suggests that the attack was made not by Ricardo but by Ricardo-Heyst. Wang does not intervene when he overhears the scuffle in the bedroom but goes outside – the discreet action of a dutiful servant. Later, waiting upon Lena and Heyst, he observes that Lena looks as if "she had been wrestling with a spirit," and Conrad observes that Wang had long regarded Heyst as "being in some sort bewitched" (p. 377). After the meal when Wang points to the bedroom and says, "Two," Heyst believes he means Heyst and Lena. These pieces can be put together in such a way that one reads from them that Lena has been wrestling with an evil spirit in Heyst. And I believe that Conrad intends by a deliberate flawing of his text, by minor inconsistencies and some illogical juxtapositions, to lead the percipient reader to that conclusion. The reader is helped along immensely by the fact that Conrad represents Heyst as being sexually aggressive. On an earlier occasion, Heyst did not obey Lena's command "to leave her alone" (p. 264).

Explorations of the kind made above lead to the question, what is the novel about? Probably most readers and critics have believed it is about an attempted robbery by three desperate men of a recluse living with his mistress on a remote island. I believe that it is about a deteriorating love relationship. It is necessary at this point to trace that relationship in some detail, insofar as it pertains to Ricardo, in the interest of estimating further the usefulness to Conrad of contemporary English psychology.

Conrad gives the reader an ambivalent impression of Heyst and Lena's life on Samburan. It has explicit gratifications for both. Lena reflects that Heyst "had opened to her the feelings of delicate joy" (p. 303) and, again, Lena "gave him a stealthy glance of passionate appreciation" (p. 239). In context these statements have a sexual meaning. Heyst's response to Lena is profound enough to erode his lifelong detachment: "the skeptical carelessness which had accompanied every one of his attempts at action, like a secret reserve of his soul, fell away from him. He no longer belonged to himself" (p. 301). Their reactions differ, however, in this way: Lena is satisfied or is potentially satisfied with the relationship, but Heyst is not; and because he is not, Lena fears that he will leave her. She develops a sense of inadequacy. It is worth sampling some passages for exact shades of meaning. Heyst's mood on Samburan at the beginning of Section III of the novel is one of "grim doubt" (p. 228) concerning the outcome of the affair, but such moments of doubt, it is said, come to him now only when he is alone. It is said, also, that he is not "accustomed" to Lena, "with whom he did not yet know how to live" (p. 245). And just before the arrival of the strangers Heyst has "a physical and moral sense of the imperfection of their relations" (p. 273). "Physical imperfection" is plain enough. The relationship is marred by a sexual disharmony attributable primarily to Heyst. The novel is not silent concerning his inadequacy.

From Heyst's own observations, one garners especially his feeling that Lena eludes him. In their "intimacy" he has a sense of "something inexplicable reposing within her" (p. 236). Because of her silences he "used to come out of her very arms with the feeling of a baffled man" (p. 398). The relationship leaves him with a "sense of incompleteness not altogether overcome – which, it seemed, nothing ever would overcome" (pp. 260-261). Heyst exclaims on one occasion, "It is because I think so much of this possession of mine that I can't have it complete enough" (p. 258). These passages do not cause the reader to conclude that Lena is at fault – although probably Conrad intends her attitudes to have some ambiguity – but rather that Heyst's requirements are inordinate – he "suffered from a trouble with which she had nothing to do" (p. 304). He is like Captain Anthony in *Chance,* to whom it occurs that he can possess the isolated Flora more completely than any other woman; he is like Razumov in *Under Western Eyes,* who plans "to steal the soul" of Natalie Haldin. Captain Anthony suffers from exacerbated passion; and Razumov's desire, in which love and a murderous intention combine, is excessive and radically abnormal. Heyst's life-long detachment, like the sexual detachment of these other figures, has left him with an unappeasable appetite according to both degree and kind.

Lena's reactions say something about the latter. In that same part of the text in which Heyst observes that he does not yet know how to live with Lena, she feels "in her innermost depths an irresistible desire to give herself up to him more completely, by some act of absolute sacrifice" (p. 247). She utters the classical female admonition to the sexually obsessed male "You should try to love me" (p. 271). At the time Heyst and Lena attempt to escape the desperados by hunting out Wang, Lena characterizes Heyst as "him whom she could never hope to understand, and whom she was afraid she could never satisfy; as if her passion were of a hopelessly lower quality, unable to appease some exalted and delicate desire of his superior soul" (p. 405). At that time the novel speaks of her "devoted, concentrated passion, burning with the hope of a more perfect satisfaction" (pp. 419-420). Probably the greatest puzzle in the novel – and there are many – is that Conrad continues this same idiom of sexual relationship in Lena's summary of the melodramatic events of the climax. As she is dying, Lena in Heyst's presence thanks God "for having been able to do it – for giving you to me in that way" (p. 495). And at the time Heyst talks about Lena's other attitude of "shame," which had its "charm." These lines reduce at least half the narrative to symbolism. What Lena has done, apart from allowing Ricardo to kiss her foot, is to coax Ricardo into giving her his knife. He reflects at the time that she is "deadly" and he "disarmed." But Lena says, "for giving you to me in that way," meaning Heyst, not Ricardo. The passage seems conclusive evidence that Conrad means Ricardo to be an aspect of Heyst. Passages assembled in the way they are above give up more easily certain meanings in the text. The novel intimates that familiar sexual aberrations of fetishism, of sadism, of the exchange of sexual roles introduce themselves into this desperate relationship of solitary lovers.

Conrad takes some suggestions for Heyst's disability, one judges, from Pessimistic philosophy. There is no mistaking the Schopenhauerian quality of his generalizations concerning "incompleteness"; he talks about the "fatal imperfection of all the gifts of life, which makes of them a delusion and a snare" (p. 261). He is a rueful example of the truth of his father's ethic, which is Schopenhauerian. Conrad got his models from real life and from close at hand. Ford Madox Hueffer's father, an immigrant musician of note in London, was an exponent of Schopenhauer's philosphy. The fact that Heyst is thought to be a Swedish nobleman reflects the Hueffer family's claim to a German barony. Ford Madox Hueffer's difficult relations with his wife and with Violet Hunt must have furnished some large part of Conrad's subject. Conrad borrowed money from Hueffer even as Morrison accepted aid from Heyst.[6] Conrad and Hueffer's collaboration may well have suggested Morrison and Heyst's common enterprise. But to return to the subject of Pessimistic influence, one can phrase easily in Schopenhauerian terms the change wrought in Heyst by his attachment to Lena. It is a painful transition, the reverse of what Schopenhauer recommends, from the world as Idea to the world as Will. Again, there is no mistaking the Schopenhauerian quality of this passage which describes the awakening of Heyst's affection while he and Lena are still in Schomberg's hotel: "he had been used to think clearly and sometimes even profoundly, seeing life outside the flattering optical delusion of everlasting hope, of conventional self-deceptions, of an ever-expected happiness. But now he was troubled; a light veil seemed to hang before his mental vision" (p. 103). Later on Samburan, when he learns of Schomberg's calumny, Heyst almost spits. The novel observes that Heyst "was not used to receive his intellectual impressions in that way – reflected in movements of carnal emotion" (p. 268). The passages say the same thing but in different ways. The first is laden with pessimistic assumptions; the second, with its reliance upon exact categories of the affective and intellectual life, is objective and almost crisply analytical. The first suggests German philosophy; the second, English psychology. And one decides that the latter ultimately is more important to the conception of Heyst's disability.

English psychologists were highly sensitive to the fact that feeling "lords it over the *intellectual* trains," to use a phrase Alexander Bain uses in *The Emotions and the Will*[7]. Awareness of themselves as nineteenth-century scientists made them so. Bain

6. See Bernard C. Meyer, *Joseph Conrad (Princeton: Princeton University* Press, 1967), pp. 132ff.

7. (New York: D. Appleton, 1888), p. 20.

observes that in the pre-scientific era feeling interfered with the observation of nature (p. 22). A generalization with which Théodule Ribot's *The Psychology of the Emotions* begins is close to Conrad's second analysis quoted above and makes it more intelligible: "emotion is in the order of feeling the equivalent of perception in the intellectual order" (p. 12). English psychologists discuss sexual abnormalities. In *The Pathology of Mind* (pp. 345-346) Henry Maudsley takes note of the shock "of sexual intercourse upon an unstable nervous temperament" under the heading of disorders ensuing from obliteration or impairment of cerebral reflexes, a classification with which Conrad's use of mobility of mood is wholly consistent. Maudsley asserts, further, that melancholy and delusions of persecution and suspicion (p. 405) result from an inadequacy for marital relationships. Heyst suffers obviously from melancholy, and his quite unexpected vehemence when he learns from Lena and Jones of Schomberg's calumny must echo a psychological analysis like Maudsley's. Certainly his reaction is out of character except as a morbid symptom. The middle-aged Heyst, schooled from his earliest years in a philosophy emphasizing human shortcomings, ought not to be outraged or even surprised by the most ordinary kind of slander.

And, finally, one goes to an aspect of the symbolism for one other psychological principle. The whole symbolism in discussion above has acquired some additional meanings which should be summarized at this point. The love relationship of *Victory* deteriorates primarily because of the temperament of its middle-aged lover. Two crises in that worsening relationship pattern the novel. The arrival of three pirates intent on stealing Heyst's treasure represents symbolically the first crisis. Conrad uses here the female-riches symbolism found in *Lord Jim.* Ricardo calls Lena a "treasure" (p. 486) even as Jewel inspires native legends of Jim's wealth. What threatens to take Lena from Heyst – and does eventually – is a "lack of equilibrium between the superior moral faculties, will and conscience, and the inferior faculties, the instincts, passions, and desires," to quote from Ribot's discussion of mobility of mood.[8] The love relationship now encompasses serious potentialities or instances of brutishness and rapacity –

8. *The Diseases of the Will* (Chicago: The Open Court Publishing Co., 1896), p. 88.

and, also, of impotence, one judges from Wang's disappearance with Heyst's revolver soon after the bandits arrive. In the novel's own terms Wang represents a firm grasp of reality, a direct gratification of desire, in short, an equilibrium between conscience and instinct. The gun is a phallic symbol. The lovers seek safety by finding and bargaining with Wang, but they fail. The second crisis occurs immediately afterwards when the bandits make their attempt on Heyst and Lena and Lena seeks to appease Heyst's or Ricardo's rapacity by some means. Conrad weighs carefully the symbolic actions and dispositions of his characters. One can believe that when he separates Pedro from Ricardo on that eventful night – Heyst insists the ape be sent away and so leaves Lena to the tiger – Conrad represents in reverse the kind of maxim Maudsley phrases as follows: the insane lack "a good sound animality to hold in solitary check the vagaries of an over-sensitive and unstable nervous system" (p. 62). The disposition of characters suggests an intensification of cruelty and mental disorder.

II

Discussion thus far has demonstrated the importance of English psychology in the period to a reading of *Victory*.[9] It should not be allowed, however, to assume overriding importance. *Victory* is predominantly a Nietzschean work with a Nietzschean psychology harmonized with the other, which antedates the Nietzschean in Conrad's fiction. Philosophical opposition to Schopenhauer[10] in the early years of the century is likely to be Nietzschean, and *Victory* is written explicitly against the admonitions of Heyst's Schopenhauerian father. The moral at the end, "woe to the man whose heart has not learned while young to hope, to love – and to put its trust in life" (p. 502), should be read, again, as an equivalent in fiction of Nietzsche's polemic against the ascetic ideal. This section will attempt to pick out those Nietzschean

9. My book *Joseph Conrad: The Making of a Moralist* (Amsterdam: Editions Rodopi, 1972) considers Conrad's use of English psychology at much greater length.

10. James Sully's *Pessimism* (London: H.S. King, 1877) attacks Pessimism from the viewpoint of English psychology.

aspects of *Victory* which are most useful in deciphering the allegory.

John A. Palmer, who has offered the most perceptive reading of it thus far, has pointed out "an evolutionary continuum among the three villains of the novel."[11] The phrase is correct, but the continuum is not only Darwinian and Spencerian, as the choice of words implies; it is also Nietzschean, and the Nietzschean is an important addition. According to the novel's descriptive terms, Pedro's aspect is "simian"; he is an "ape"; but he is also a "bear" (e.g., p. 128). As he chooses metaphors in this instance, Conrad's mind seems poised between the Darwinian and the Nietzschean. For the animal Nietzsche perceives in man may be an ape, but it is more characterisitically a performing bear.

Ricardo owes more to Nietzsche than to English evolutionary literature. He is the "tiger" English psychologists describe, feral and predatory. But, more importantly, he is Nietzsche's "beast of prey"; and Conrad uses those words repeatedly to describe him. The contemptuous contrast Ricardo draws between himself and Schomberg brings out the unmistakably Nietzschean quality of both characterizations. Schomberg is "tame" and a wage "slave" (p. 179). Ricardo is neither. Nietzsche's slave in antiquity suffers from punishments inflicted by cruel masters, who are beasts of prey. Prohibitions and "sinister mnemonics" tame the animal turned man and cause him to internalize primordial instincts. Made ashamed of his animal nature, man creates for himself a conscience and a soul. The slave's penchant for cruelty and his will to power are suppressed without being eradicated. The slaves become good haters, filled with a festering venom and malignity. To relieve man's suffering soul, man invents religion. The priest prescribes cautious doses of spiritualized or deified cruelty, as in the festive spectacle of the *auto-da-fé*. Slave morality, which recommends virtues opposite to those of the masters – altruism, meekness, restraint, asceticism – effects a "revolt of the slaves" in the form of a "transvaluation of values." Schomberg is Conrad's conception of the German temperament which provided much raw material for Nietzschean psychology. Not only is he "perfectly tame" and of a slave temperament, he is also venemous

11. *Joseph Conrad's Fiction* (Ithaca: Cornell University Press, 1968), p. 192.

and malign. Nietzsche's masters become angry, but his slaves hate. Conrad draws a Nietzschean contrast when he describes Schomberg's "lurid, heavy, Teutonic manner, so unlike the picturesque, lively rage of the Latin races" (p. 131). And Conrad elaborates the same Nietzschean analysis in Schomberg's calumny, which is so damaging to the over-sensitive Heyst. If Ricardo is not a slave like Schomberg, he is also not a master; but he serves one and so enjoys the freedom of the masters. Conrad conceives of Ricardo as a follower; Ricardo says to Schomberg, "I ain't a dog walking on its hind legs for a bone; I am a man who's following a gentleman" (p. 179). Jones and Ricardo's ethic and way of life are written according to Nietzsche: to Ricardo "life was not a matter of passive renunciation, but of a particularly active warfare" (p. 319). Jones "depended on himself, as if the world were still one great, wild jungle without law" (p. 141). For the masters, who, according to Nietzsche, "come like Fate" and have nothing to do with contracts,[12] life is unpredictable. Conrad turns that Nietzschean characteristic into fiction, one judges, playfully and with some irony, by making Jones a gambler. Jones may not strike the reader as a master even yet; but Conrad seems to think of him in terms of Nietzsche's remarks on the criminal temperament: "The criminal type is the type of the strong man amid unfavourable conditions, a strong man made sick" (XVI, 103). The contrast between Ricardo and Schomberg and the Ricardo-Jones relationship cannot be extracted from English evolutionary literature and are in excess of an "evolutionary continuum."

The fullness of the master-follower relationship, which is skillfully drawn and inserted in the novel, is, nonetheless, somewhat gratuitous and somewhat falsifying with reference to symbolic meaning, where the important relationship of Ricardo and Jones is the relationship between beast-of-prey cruelty and what is made of it historically. It is in Nietzsche's words "by frightful mutilations and chronic depressions" turned into its opposite (XIII, 186). The novel emphasizes Ricardo and Jones' temperamental similarity: "one the outcast of his vices, the other inspired by a spirit of scornful defiance, the aggressiveness of a beast of prey looking upon all the tame creatures of the earth as its natural

12. *The Complete Works of Friedrich Nietzsche* (London: T.N. Foulis, 1909-13), XIII, 103.

victims" (p. 331). Ricardo's pleasure is the growl and the spring that liberates "the suffering soul of savagery" (p. 354). Jones' pleasure is toying with Heyst. One is the mutilation and sublimation of the other. Jones in his thinness, his illness, his boredom, and homosexuality is what Nietzsche sees in modern man when he concludes that the blonde beast is preferable to "that excess of deformity, disease, exhaustion, and effeteness whose odour is beginning to pollute present-day Europe" (XIII,43). Again, one cannot extract this figure from the writing of English evolutionists, who believe in progress. One must go to the German "pessimists."

One concludes, finally, that if Ricardo and Heyst are the same person, Jones and Heyst are also; or that all three bandits are aspects or potentialities of Heyst's temperament. Awareness of the Nietzschean quality of Ricardo and Jones opens up that insight to the reader inasmuch as it establishes also the identity of Ricardo and Jones. The novel ties Heyst and Jones together in other ways. They are both gentlemen in Ricardo's master-follower idiom. Jones, further, insists on their similarity: "You and I have much more in common than you think" (p. 394). He claims that they pursue the same ends; and that is a Nietzschean assertion, for it is true in the Nietzschean sense that both he and Heyst seek to live outside the strictures of a slave morality. Heyst's father in his "Look on, make no sound" advice somewhat revises Schopenhauer. Another passage quoted from Heyst's father ties the two together: "Men of tormented conscience, or of a criminal imagination, are aware of much that minds of a peaceful, resigned cast do not even suspect" (p. 269). "Tormented conscience" and "resigned cast" are phrases of Nietzschean analysis; and the comparison is inspired obviously by Nietzsche's depiction of the criminal temperament, an example of which has been quoted already. If Heyst and Jones are one, Heyst's "fastidiousness" (p. 106) with respect to Lena matches Jones' horror of women and says something about the nature of the difficulties of the sexual relationship. The alienation of Jones and Ricardo because Ricardo is attracted to Lena conveys successfully a sense of psychic change as well as a sense of a perilous fragmentation of mind or, to fall back upon the English psychologists, a disequilibrium between higher moral faculties (the novel, p. 326, describes Jones' attitude toward women as a kind of "depraved morality") and basic instincts and desires. To relieve that pain and to rid himself of "disgust," Heyst or Jones shoots Lena, aiming at Ricardo, thus killing the cause; and in time he will kill the effect. The title of the novel is ironic. No one gains a victory.

The other deaths do not seem as meaningful. Jones and Pedro die when, it may be thought, Heyst is returned to a normal hold on reality. Wang's role in shooting Pedro has that significance. The Chinaman in English psychology is an example of the disciplined temperament perfectly accommodated to an advanced state of civilized life; and he has much the same favorable significance in Nietzsche. Conrad seems to echo both. When the strangers first arrive and are being provided for, Wang hands Pedro a kettle "at arm's length, as if across a chasm" (p. 300). The chasm is all the centuries of human evolution. Conrad depicts Wang also in a Nietzschean way as free from illusion and false idealistic sentiment. He alludes to Wang's "simple feeling for self-preservation, untrammelled by any notions of romantic honour or tender conscience" (p. 376).

The Nietzschean aspects of *Victory* not so immediately relevant to the symbolism as the ones above should be considered before the subject of Nietzschean influence is abandoned. Lena's passivity, *i.e.*, her exhaustion, her hours of sleep, after encounters with Ricardo-Heyst – a dominant impression – like the stillness and immobility of Flora de Barral in *Chance* on the occasion of the governess' savage attack, must have been suggested by that "enervation" which is the characteristic reaction of Nietzsche's victims and slaves. Secondly, fatality and causality are of as much interest to Conrad in this novel as they are in *Chance* and *Under Western Eyes.* The two earlier novels exhibit an almost Hardian interest in the concatenations of external circumstance, and the hand of Providence is never entirely obliterated even when fatalistic sequences are traced naturalistically to aberrations of a Nietzschean kind, such as Captain Anthony's idealistic chivalry and Razumov's spiritualized cruelty. As has been noted, *Victory* contains a contrast between the ordered and predictable life of a highly organized, essentially democratic society and the gambler's freedom of a world composed of masters, followers, and slaves; but that Nietzschean contrast is elaborated at much greater length in *Under Western Eyes.* Conrad's interest in *Victory* turns not toward fatalistic sequential effects but inward in the direction of causality. Jones describes himself to Heyst as a "sort of fate – the retribution that waits its time" (p. 464), and Ricardo seems to Lena to be "the embodied evil of the world" (p. 366). Conrad in these passages again attributes personal and social disasters, society's uneasiness and unrest, to human complexity as Nietzsche depicts it. What is new in *Victory* is that Conrad's interest in Providence also turns inward in a sense. With reference to being at the mercy of the three bandits, Lena asks whether it is not "a sort of punishment" (p. 432). And Heyst answers, "Are our visitors then messengers of morality, avengers of righteousness, agents of Providence? That's certainly an original view" (p. 433). These comments in the whole context of the novel are tantamount to saying that Nature harmonizes and restores. It works inwardly to impel the individual to involve himself in society. It remedies the maladies of decadence – Heyst is its one conspicuous failure – by strengthening instinctive life. Conrad gives a transcendental coloring to what is essentially a Nietzschean ethic. It is a mark in him of nineteenth-century optimism, intellectually speaking; but the ultimate source, as suggested before, is a pleasure of Blakean intensity in the experience of life.

Finally, one credits Nietzsche with stimulating Conrad's pictorial imagination. Conrad describes Heyst's parleying with Jones on the night of the storm as "the conversation of an evil spectre with a disarmed man, watched by an ape" (p. 351). The three marauders on another occasion are "evil intelligence, instinctive savagery, arm in arm. The brute force is at the back" (p. 403). Conrad acknowledges sufficiently what he is doing when he calls them "fantasms from the sea – apparitions, chimeras! And they persist" (p. 404). Similar materializations and the same mixture of the mythical and the real exist in passages of Nietzsche, which Conrad's epithets recall. Especially relevant are those Nietzschean freaks, genius bound to the "billy goat" and the "ape" and the "scientific head" on "an ape's body" (XII,39). Recent criticism has called attention to the presence of the mythical in *Victory*,[13] but it has not discovered that the mythical exists there in a very

13. See Mary B. Goens, "The 'Mysterious and Effective Star': The Mythic World-View in Conrad's *Victory*," *Modern Fiction Studies*, XIII (1967), 455-63.

special form, as the basis for allegory rather than as spontaneous symbol produced by the unconscious imagination. However, allegorical figures require a suitable setting, which Conrad gives them by freely recreating the Eastern world of his early adventures. It is a world with a round island and unimaginable spatial reaches, with ravishing contrasts of sun and shade and storms dramatically timed. It is more than Polynesia. It is the psychic landscape of a north European who once enjoyed a profound intimacy with the tropics.

III

Discussion thus far has demonstrated Conrad's weight as a moralist in *Victory,* and it has exposed his finished quality in considerable detail. Judged by a standard of this kind, *Victory* is the greatest of Conrad's novels. But it has been held in disesteem not only because of content but also because of what have been regarded as its technical shortcomings. Adverse judgments have usually failed to take into account the exact nature of the work. John Palmer's *Joseph Conrad's Fiction* describes *Victory* accurately as a mixture of realism and allegory. Palmer hypothesizes that these levels "interfere" with each other (p. 182). He overlooks the fact that some action as well as some characterization is allegorical. The result is not strictly *interference* but *ellipsis* in the realistic narrative. Lena's thanking God for giving Heyst to her "in that way" is the novel's most notable example. Earlier she despairs of being able to "satisfy" Heyst's desire. That aspect of the unsatisfactory relationship of the lovers follows the arrival of the bandits and Wang's disappearance with Heyst's pistol, but it depends upon no literal representations in the text. The realistic narrative breaks off; and when it resumes, there occurs what might be called an overloading of the text because the realistic narrative relies upon the meaning of the intermittent symbolism. This phenomenon occurs in most allegorical literature from *The Faerie Queene* to *Passage to India,* and that fact must be Conrad's defense. But one can still separate what is obscurantist in *Victory* from what is characteristic of the genre. Conrad was not a Naturalistic writer, and he veiled his Naturalistic subject to the general reader in 1914. Conrad wrote into his novel much that was personal and, it seems, much that was scandalous; and that too he disguised. A blemish there is in *Victory*, but it is a small blemish upon a magnificent novel.

IV

The section that follows is a postscript on Conrad's use in *Victory* of classical mythology. Conrad's adaptation of the Acis-Galatea myth confirms, I believe, that his view of his subject is Nietzschean. Near the novel's climax Jones, observing Lena and Ricardo in Heyst's bungalow, draws this comparison: "Behold the simple Acis kissing the sandals of the nymph, on the way to her lips, all forgetful, while the menacing fife of Polyphemus already sounds close at hand – if he could only hear it! "(p. 482). The classical parallel, although not prominent elsewhere, influences, I believe, certain aspects of setting, characterization, and action and is important to theme.

The Eastern setting of *Victory*, a partially bucolic landscape, with sun and shadows among spaced trees, differs from the sinister jungle which is the Eastern setting of Conrad's early fiction. One can imagine himself in Sicily, legendary home of the Cyclops. The impression is strongest in those interludes of love in which Heyst and Lena leave the clearing and climb the heights of Samburan. It is as though an unconscious association of Heyst and Polyphemus dallying with the sea nymph has drawn from Conrad's imagination the more appropriate setting. Further, once the classical allusion is acknowledged, vestiges of that ruinous scheme Heyst once involved himself in, *i.e.*, the Tropical Belt Coal Company – a signboard and some huts called "bones" (p. 211) – strike one as a playful, somewhat distant parallel to the remains of the Cyclops' victims.

Homer in the *Odyssey* describes both the island of the Cyclops and another lying athwart its harbor.[14] In *Victory* Conrad makes much of an island volcano visible across the bay from Heyst's bungalow; Davidson remarks that thunderstorms gather there (p. 501). Conrad's volcano is not Homer's "level" island; but recollections of classical accounts of the islands of Lipara and Strongyle and of Sicily, where Vulcan had his workship and with

14. Trans. A.T. Murray (Cambridge: Harvard University Press, 1960), I, 311.

the help of the Cyclops fashioned thunderbolts for Zeus,[15] may have caused Conrad to alter Homer's topography. On the night of Lena's death the thunder is like the growl of an "inarticulate giant" (p. 473). Conrad compares the volcano glowing in the dark and Heyst smoking (p. 4). This "volcanic" association may have resulted from the metaphor Ovid in the *Metamorphoses* puts in the mouth of the lovelorn Polyphemus "I seem to carry Aetna in my breast."[16]

Conrad, seemingly, stages visual effects and employs some verbal tricks to reproduce in Lena and Heyst the disproportionate sizes of Polyphemus and Galatea. Immediately after drawing the classical comparison quoted above, Jones commands Heyst, "Stoop a little." Earlier Heyst, approaching his veranda, sees Lena only from the waist down, an effect of the lantern's light (p. 302); but, also, it is as though from a considerable height, the roof is a partial screen. Heyst several times lifts Lena, as though she is a child, into his embrace, the strongest such effect occurring in a love episode on the heights of Samburan: "till then she had never felt herself swinging between the abysses of earth and heaven in the hollow of his arm. What if he should grow weary of the burden! " (pp. 256 -57).

Classical description may be thought to influence some other physical details. Heyst possesses great strength. He is also attractive, a fact that does not agree with descriptions of Polyphemus in the epics. But the Cyclops is more pleasing in Theocritus and Ovid and in Virgil's Eclogue II. Beard, eye, and teeth figure in these writers and have near equivalents in Heyst's grey eye, martial moustaches, and teeth which flash agreeably. Lena's skin is dazzlingly white (p. 497) like Galatea's in the bucolic poets. Lena has a "sculptural forehead" (p. 309) and a goddess' stance (p. 310). One passage in *Victory* applies to her a metaphor suited to a Nereid returning to her element. Lena "was surprised by a wave of languid weakness that came over her, embracing and enveloping her like warm water, with a noise in her ears as of a breaking sea" (p. 378).

15. See William King, *An Historical Account of the Heathen Gods and Heroes* (Carbondale: Southern Illinois University Press, 1965), pp. 130-31.

16. Trans. Frank Justus Miller (London: William Heinemann, 1926), II, 289.

Two versions of Polyphemus' love for Galatea exist in classical accounts. In the more famous, *i.e.*, in Ovid's *Metamorphoses* (II, 281ff.), Galatea spurns Polyphemus' affection as she laments to the Nereids the death of her lover Acis. Appian's *Roman History*, in contrast, calls Galatea Polyphemus' wife.[17] And Theocritus' Idyll VI pictures a practiced Polyphemus, "fair i' the eyes of love," who pretends indifference as he is pursued by an aggressive, wanton, and jealous Galatea.[18] Conrad surely echoes this legend when Lena, distressed by Heyst's mental abstraction and complexity, seeks to possess him more completely. Conrad, one judges, exploits all of the Cyclopean myths. In the novel's climactic scene, shortly before Ricardo dies, Conrad appears to paraphrase the death of Acis, who was turned into a river with which the Nereid mingled: "I went tired this morning, since I came in here and started talking to you – as tired as if I had been pouring my life-blood here on these planks for you to dabble your white feet in" (p. 486). Conrad uses, further, the blinding of Polyphemus as told not in the *Odyssey* so much as in Theocritus' *Idyll* XI.[19] The candles by whose light Heyst observes Ricardo with Lena on the night of the storm "flung around her an intolerable brilliance which hurt his eyes, seemed to sear his very brain with the radiation of infernal heat. It was some time before his scorched eyes made out Ricardo seated on the floor" (p. 481). *Idyll* XI contains a suggestion for equating, in the fashion of the passage above, the blinding of the Cyclops with love for Galatea. Polyphemus says there that Galatea may burn his soul and all his possessions, even the dearest, his one eye.[20] Finally, Conrad's use of the sun in the novel recalls the Cyclops' death at the hands of Apollo. The sun is hostile throughout, "all aquiver with the effort to set the earth on fire, to burn it to ashes" (p. 374). One thinks back naturally to this passage when faced with Heyst's rather mysterious immolation. At the novel's close the myth is not explicit, but the mythical seems to determine the unnaturalistic character of the naturalistic narrative.

17. Trans. Horace White (New York: Macmillan, 1899), I, 258.

18. See *The Greek Bucolic Poets*, trans. J.M. Edmonds (London: William Heinemann, 1919), pp. 85 -89.

19. See *ibid.*, pp. 141 -47.

20. *Ibid.*, p. 145.

Victory is not Joyce's *Ulysses* with the latter's exact and extensive parallels, but similarities cited above lead to the conclusion that the legend is more than a rhetorical ornamentation of a small portion of Conrad's narrative.

A passage from an article in the *English Review* by Frank Harris gives forthrightly a meaning the legend had for Conrad's contemporaries. Harris alters the amorous preference for boys Polyphemus has in classical myths, adapting it to marital inadequacy such as Thomas Carlyle experienced.

> Carlyle's impotence made everything about him clear to me. Ever afterwards I saw him as a sort of Polyphemus, a one-eyed giant: he stood to me for Puritanism itself and explained it, in its strength and in its fatal weakness as no one else could. Paganism died because it neglected the soul, and the claims of the soul; Puritanism died because it scorned the body and the claims of the body.[21]

In a more personal vein Harris claims that Carlyle gave him the whole history of his married life: "In brief the story was that he admired his wife beyond all other women, loved her and her alone all his life, but had never consummated the marriage or lived with her as a wife" (p. 432). This aspect of Carlyle's temperament, no doubt, is one important reason for Ford's – and seemingly Conrad's – finding Carlyle an "intolerable" moralist. Harris' application of the myth is instructive with reference to Conrad's application and suggests that *Victory* is about a deteriorating sexual relationship.

Nietzsche uses the myth of the Cyclops in this passage from *Beyond Good and Evil:* "Pity has an almost ludicrous effect on a man of knowledge, like tender hands on a Cyclops" (XII, 470). In the sense that he leads a life of detachment dictated by a Schopenhauerian cynicism and asceticism, in the sense that he analyzes lucidly the change worked in him by Lena from intellectual to sensual apprehension, Heyst is "a man of knowledge." Conrad uses "compassion" often in *Victory* to describe Heyst's feeling for Lena (*e.g.*, p. 99). Pity for Nietzsche and, one judges, for Conrad in this instance is transformed instinct, self-mutilation; it is charity in the beast of prey, delicacy in the sexually ravenous; it is the ludicrous disability of Heyst and the Cyclops.

21. "Talks with Carlyle," *The English Review*, 7 (Dec. 1910-March 1911), 434.